Advances in Spatial Science

Springer
Berlin
Heidelberg
New York
Barcelona
Hongkong
London
Milan
Paris
Singapore
Tokyo

Titles in the Series

C. S. Bertuglia, M. M. Fischer and *G. Preto* (Eds.)
Technological Change, Economic Development and Space
XVI, 354 pages. 1995. ISBN 3-540-59288-1
(out of print)

H. Coccossis and *P. Nijkamp* (Eds.)
Overcoming Isolation
VIII, 272 pages. 1995. ISBN 3-540-59423-X

L. Anselin and *R. J.G.M. Florax* (Eds.)
New Directions in Spatial Econometrics
XIX, 420 pages. 1995. ISBN 3-540-60020-5
(out of print)

H. Eskelinen and *F. Snickars* (Eds.)
Competitive European Peripheries
VIII, 271 pages. 1995. ISBN 3-540-60211-9

J. C.J.M. van den Bergh, P. Nijkamp and *P. Rietveld* (Eds.)
Recent Advances in Spatial Equilibrium Modelling
VIII, 392 pages. 1996. ISBN 3-540-60708-0

P. Nijkamp, G. Pepping and *D. Banister* (Eds.)
Telematics and Transport Behaviour
XII, 227 pages. 1996. ISBN 3-540-60919-9

D. F. Batten and *C. Karlsson* (Eds.)
Infrastructure and the Complexity of Economic Development
VIII, 298 pages. 1996. ISBN 3-540-61333-1

T. Puu
Mathematical Location and Land Use Theory
IX, 294 pages. 1997. ISBN 3-540-61819-8

Y. Leung
Intelligent Spatial Decision Support Systems
XV, 470 pages. 1997. ISBN 3-540-62518-6

C. S. Bertuglia, S. Lombardo and *P. Nijkamp* (Eds.)
Innovative Behaviour in Space and Time
X, 437 pages. 1997. ISBN 3-540-62542-9

A. Nagurney and *S. Siokos*
Financial Networks
XVI, 492 pages. 1997. ISBN 3-540-63116-X

M. M. Fischer and *A. Getis* (Eds.)
Recent Developments in Spatial Analysis
X, 434 pages. 1997. ISBN 3-540-63180-1

R.H.M. Emmerink
Information and Pricing in Road Transportation
XVI, 294 pages. 1998. ISBN 3-540-64088-6

P. Rietveld and *F. Bruinsma*
Is Transport Infrastructure Effective?
XIV, 384 pages. 1998. ISBN 3-540-64542-x

P. McCann
The Economics of Industrial Location
XII, 228 pages. 1998. ISBN 3-540-64586-1

L. Lundqvist, L.-G. Mattsson and *T. J. Kim* (Eds.)
Network Infrastructure and the Urban Environment
IX, 414 pages. 1998. ISBN 3-540-64585-3

Roberta Capello · Peter Nijkamp
Gerard Pepping

Sustainable Cities and Energy Policies

In Association with:
Kostas Bithas, Roberto Camagni
Harry Coccossis

With 30 Figures
and 37 Tables

Springer

Prof. Dr. Roberta Capello
University of Molise
Department of Economics
and
Politecnico of Milan
Department of Economics
Piazza Leonardo da Vinci, 32
I-20133 Milan
Italy

Prof. Dr. Peter Nijkamp
Dr. Gerard Pepping
Free University
Department of Regional Economics
De Boelelaan 1105
NL-1081 HV Amsterdam
The Netherlands

ISBN 3-540-64805-4 Springer-Verlag Berlin Heidelberg New York

Cataloging-in-Data applied for
Die Deutsche Bibliothek - CIP-Einheitsaufnahme
Capello, Roberta:
Sustainable cities and energy policies / Roberta Capello; Peter Nijkamp; Gerard Pepping. In assoc. with: Kostas Bithas ... - Berlin; Heidelberg; New York; Barcelona; Budapest; Hong Kong; London; Milan; Paris; Singapore; Tokyo : Springer, 1999
(Advances in spatial science)
ISBN 3-540-64805-4

Hardcoverdesign: Erich Kirchner, Heidelberg
SPIN 10689181 42/2202-5 4 3 2 1 0 - Printed on acid-free paper

Preface

Our world is becoming increasingly urbanized - with all the advantages and disadvantages that go with it. The geography of the twentieth century exhibits an intensified trend towards an urban way of life in modern society. Despite sub-urbanization - and sometimes de-urbanization tendencies - the city remains the nucleus of a developed economy. It is undoubtedly true that the economies of density and scale are decisive factors for city formation. Clearly, there are also dis-economies as witnessed by congestion, environmental decay and so forth. Nevertheless, the positive features of the city still appear to be a dominant force, as the city is an extremely efficiently organized geographical entity.

This holds good in particular for the use of public utilities such as water, telecommunications, electricity and energy, in general. Also, the agglomeration advantages of a city make it possible to adopt new forms of environmental waste management and renewable energy policies which otherwise would not be feasible.

This background has led to the notion of a sustainable city, a concept which refers to the potential of urban agglomeration to ensure an environmentally benign development of a city through focused environmental and energy initiatives which stimulate a balance between economic progress, social equity and environmental quality.

This book starts out with the optimistic perspective that modern cities can indeed play a strategic role in the necessary pathway to sustainable development, with particular emphasis on the opportunities offered by local energy and environmental initiatives. Our study aims to demonstrate that an urban sustainability policy has many socio-economic benefits, while it also seeks to identify the critical success and failure factors of sustainable city innovations. After a comprehensive review of various opportunities and experiences, attention is focused particularly on renewable energy resources which may offer new potential for the active involvement of local authorities.

The study also highlights major impediments regarding the adoption and implementation of renewable energies, in particular, the development of advanced energy-environmental technology in a world dominated by natural (public) monopolies and/or monopolistic competition elements. In this context both theoretical and empirical elements are discussed, as well as institutional aspects.

The theory and methodology is tested by a thorough empirical investigation into local renewable energy initiatives in three European countries, viz. Greece, Italy and The Netherlands. Based on an extensive data base, various statistical models are estimated in order to identify the key elements and major driving forces of sustainable development at the city level. And finally, the study is concluded with a long list of applicable and operational policy guidelines for urban sustainability. These lessons are largely based on meta-analytic comparative studies of the various initiatives investigated.

This publication is the result of a study commissioned by the European Union (DG XII) which aimed to bring together and to disseminate practical knowledge about urban renewable energy policies. We wish to thank Kostas Bithas (Panthios University, Athens), Roberto Camagni (Politecnico, Milan) and Harry Coccossis (University of the Aegean, Mytilini), for having put much of their knowledge and time into the project with their creative and constructive suggestions. Patrizia Favara and Thomas Ursem offered much help in collecting the information for the empirical analysis. Our final vote of thanks goes to all experts and people we interviewed in the twelve cities investigated. Without their cooperation, our research would have not been possible. The constructive comments of Geoffrey Hewings (Illinois) are acknowledged and great help during the editorial stage was offered by Maureen Shepstone (Amsterdam) and Enrico Sabella (Bologna). Last, but not least, the 'master of ceremonies' of Springer-Verlag, Marianne Bopp, must be thanked for her successful efforts in coordinating the various players in their publication game.

Milan/Amsterdam, August 1998
Roberta Capello, Peter Nijkamp and Gerard Pepping

Table of Contents

PART A

THE SUSTAINABLE CITY

1 Concept of a Sustainable City

1.1 Introduction: Aims and Scope

Since the early 1960s, when Rachel Carson's *The Silent Spring* attracted world-wide attention, there has been an ever-increasing awareness of the extensive damage to the environment caused by various forms of pollution. An avalanche of literature has been published since the 1980s on the pervasiveness of environmental decay ranging from local to even global scales and culminating in the widely cited Brundtland Report (WCED 1987). Sustainable development has clearly taken on a global dimension, but in recent years it is has increasingly been acknowledged that there is a close mutual interaction between local and global processes. Localities (e.g. cities, villages) are open spatial economic and ecological systems impacting on their surroundings and on the earth as a whole. The recognition that much of the sustainability debate has an urban orientation is also based on the fact that cities are large consumers of natural resources and major producers of pollution and waste. For example, the cities in OECD countries consume approximately 60 to 80 percent of total energy demand (see OECD 1995). The role of localities is even more pronounced, when we recognize that cities are also the major sources of new technology, economic growth and new environmental initiatives. Consequently, the role of the city is increasingly that of an animator and coordinator of creative quality-of-life strategies. And this role is likely to be reinforced in the future (Gibbs 1994; Girardet 1992b).

Sustainable development has become one of the touchstones of urban policy in the past years. And consequently, the notion of *a sustainable city* has recently gained much popularity in many countries (see e.g. Nijkamp and Perrels 1994; Haughton and Hunter 1994; Selman 1996). There is at present considerable analytical and political interest in the success conditions for a sustainable city. The urban focus in the present sustainability debate is largely caused by the fact that (large) cities are the major users of nature's scarce resources and the major causes of environmental decay (cf. Giradet 1992a) At the same time, cities are able - as a result of scale advantages - to create more efficient energy savings conditions and related environmental quality improvements (Capello 1998). Furthermore, it ought to be recognized that the sphere of influence of cities - in terms of spatial interaction of persons, goods, environmental impacts and resource use - extends far beyond their own territory; the potential of cities to shape attractive quality of life conditions in areas under stress means that cities are sometimes regarded as *islands of opportunities in seas of decay* (cf. Van Geenhuizen and Nijkamp 1997). A currently vogue term in this context is that of *an ecological footprint*, which refers to the fact that the environmental burden of a city - through use of scarce inputs and through pollution emission - may have a formidable geographical coverage. Thus, the environmental carrying capacity of a city would ideally have to be placed in a much wider spatial setting (see Archibugi 1997; Rees 1992; Wackernagel and Rees 1996).

It has rightly been argued in the literature that urban sustainable development is not only a matter of environmental quality control. The city is essentially the result of three main intersecting forces, viz. the social, the environmental and the economic forces, which are interlinked, giving rise to positive and negative factors (see Camagni 1998). Urban sustainability goals are thus related to the maximization of positive factors stemming from the interaction of these three elements (e.g. a high quality of the labour market, increasing returns in energy use, economies of density in pollution control etc.) and the minimisation of negative factors (e.g. traffic congestion, air, water and soil pollution). Thus, sustainable cities are not only characterized by a clean environment, but have a much richer socio-economic and environmental scope.

A central element in creating urban environmental sustainability is the adoption and implementation of appropriate energy policies, since most environmental factors in cities are directly or indirectly related to urban energy use. Many cities have in the past years developed various types of sustainable city strategies. The current practice demonstrates an overwhelming variety of initiatives and policies, so that the actual success of such strategies, seen from a cross-sectional comparative perspective, is hard to assess and evaluate.

The main focus of this book is on energy use, with particular emphasis on the central role played by *renewable energy* in cities as a key element in achieving sustainable urban development. Renewable energy policies comprise a wide spectrum of options ranging from solar, thermal or wind energy to photovoltaic (PV) energy systems, waste recycling activities and combined heat and power (CHP) applications. Renewable energy provision may offer a significant contribution to an improvement of local or regional environmental quality conditions. As a result, new policy initiatives focusing on the intricate relationship between renewable energy alternatives and local sustainability conditions are no doubt warranted. However, such policies sometimes meet much resistance, so that their economic, environmental and technological potential is not fully exploited. Therefore, there is a need for a more extensive and in-depth analysis of the critical success factors for the adoption and use of environmentally-benign local energy initiatives.

The aim of this book is to highlight the great potential of decentralized (i.e. local or urban) energy policies in achieving environmentally-benign developments for modern cities. Urban sustainability will be placed in the context of the debate on global sustainable development. A wide array of policy initiatives will be discussed and evaluated, ranging from market-based energy policies to technological innovation policies for the energy sector (Part A). A theoretical framework for technology adoption processes will be developed and empirically tested (Part B). The main question addressed is: which are the critical success factors for successful urban energy policies? This question will also be dealt with in a meta-analytic context by assessing and comparing the performance of energy policies in various European cities, with a particular view to renewable energy (Part C). Conclusion and policy guidelines are offered in Part D.

In this introductory chapter, we will first focus on the more general concept of urban sustainability, as decentralized energy reforms must be seen as part of overall urban environmental programmes.

1.2 Pathways to Sustainability

Modern cities are in a state of flux. Economic restructuring, socio-political transformation, technological innovation and global networking have exerted profound impacts on the position of cities and regions in our world. Cities tend to become both centrifugal and centripetal nodes in a national - and also increasingly international - society linked by means of networks. This also holds for cities in developing countries; they face the urgent need to solve problems of poverty, housing and unemployment, and at the same time they are also facing the need to generate competitive advantages in order to be sustainable and to survive in a fierce national or international competition (see Nijkamp 1991; Porter 1990). Against the background of spatial competitiveness based on competition and gains or losses of trade (of goods and services), the idea of spatial sustainable development is increasingly coming to the fore.

The notion of *sustainable development* has been discussed extensively in the past decade (see e.g. WCED 1987; Archibugi and Nijkamp 1989; Opschoor 1992; Van den Bergh 1996). However, many discussions have mainly addressed sustainability on a global scale, and to a large extent they focused on conceptual and measurement issues. Consequently, it is not surprising that the lack of empirical applicability of this concept in a concrete policy context has been a major source of criticism. A more operational treatment of the rather abstract concept of sustainable development seems to be necessary in the current debate on our common future, as was also strongly emphasized at international fora held in Rio de Janeiro (1992) and Kyoto (1997).

Sustainable development has been defined as (WCED 1987): "*... a pattern of development that meets the needs of present generations without jeopardizing the ability of future generations to meet their own needs*". This can be interpreted in the sense that economic development should leave its resource base intact, in order to offer future generations equal access to possibilities to use the natural environment as an input into the process of welfare generation (Opschoor 1992; James *et al.* 1989). This resource base refers to two elements: the physical resources themselves, i.e. the total of all relevant environmental stocks (such as fisheries, forests, arable land, water, air quality, etc.), and the 'environmental infrastructure', i.e. the environmental processes and life support systems (e.g. resource regeneration processes and waste absorption systems).

Sustainable development does not only have an intergenerational dimension, but it also contains other aspects, notably spatial allocation aspects (cf. Barrow 1995). Such spatial dimensions range from global to local. Global examples include: sea level rise of the oceans mainly caused by global warming or climatic changes; depletion of the ozone layer; deforestation; desertification; and extinction of species. Local examples relate to e.g. pollution in areas that face strong urban development; hazards from chemical, nuclear and radioactive toxic industries; road construction, or local catastrophes. A common attribute of these examples is that they threaten human and environmental survival and hence are 'negative' or 'undesirable' changes that one should try to avoid or otherwise cope with. 'Positive' changes (regeneration of natural environmental stocks, advances in environmental technology etc.) are by definition 'welcome' and thus beneficial to human welfare. Intensified scientific action is needed to support and enhance such 'positive' changes.

While some early research seemed to view sustainable development as an end-point, it has been argued more recently that sustainable development is better conceived as a process or even an organizing principle. Whether this is any more use to practitioners attempting to implement policy initiatives on the ground is debatable. The lack of more clear and more specific definitions of sustainable development for practical application than those contained in the WCED (1987) report still remains the weak side of the continuing debate on sustainability policies. Van den Bergh and Van der Straaten (1994) argue that much work emanating from sustainable development does not seem to go beyond abstract or descriptive studies, with the consequence that many scientists are becoming somewhat cautious about using the term. Nevertheless, sustainable development has increasingly pervaded policy-making at all geographical scales ranging from the international to the local.

Clearly, an operational definition of global sustainability is almost impossible to give. But at a meso level many more appealing possibilities do exist, e.g. by using measurable criteria such as carrying capacity, critical thresholds, maximum yield etc. Thus, the use of clear environmental quality indicators and of threshold values seems to offer a viable pathway to a meaningful application of sustainability concepts in policy practice (see also Giaoutzi and Nijkamp 1994, and Nijkamp and Ouwersloot 1998). The meso level of attention is of course also interesting from an urban sustainability perspective.

The Local Agenda 21, arising from the UN Conference on Environment and Development in Rio de Janeiro, has put particular emphasis on the latter issue (UNCED 1992; Whittaker 1995). Local Agenda 21 has both set the agenda across a large spectrum of policy and management issues for the twenty-first century and has more broadly assessed the acceptance of sustainability based on the inseparability of social, economic and environmental issues. This agenda has also called attention to three key environmental concerns, namely climate change, biological diversity and forest management, but their actual policy relevance was not unambiguous. Some recent contributions in the area of sustainable development have tried to provide more concrete guidance and practical examples which are of use for both academics and practitioners in order to come to grips with sustainable development, by addressing also issues of policy and implementation; see e.g. Van den Bergh and Van der Straaten (1994), Nijkamp and Perrels (1994) and Haughton and Hunter (1994). The last two mentioned authors also put much emphasis on the social context for the acceptance of sustainable development. Another example of an attempt to provide more concrete guidelines to sustainable development is found in Van Geenhuizen and Nijkamp (1997), who have tried to operationalize an integrated planning approach for sustainable development, by focusing specifically on empirical needs and data architectures in an urban policy context.

In the next section we will address more specifically some conceptual elements of spatial (i.e. local) sustainable development, while in subsequent sections we will highlight the promising role of cities in this context.

1.3 Local Sustainable Development

Within the general context of sustainability, it makes sense to analyse and manage socio-economic and environmental processes at a 'disaggregate', i.e. local level. The inherent logic of a local or urban sustainable development analysis is based on the fact that such an analysis may make functional interdependencies at the local level more manageable, seen from the viewpoint of the long run development of cities and regions. From a management and policy point of view, the local level is more suitable for policy control and transformation than the global scale. And hence it is evident that the objective of sustainable development may be achieved more easily, if the processes of socio-economic development and environmental change at a local scale are more clearly understood and properly managed (see also Nijkamp and Perrels 1994).

To achieve sustainable development, an integrated economic and environmental approach to local level policy-making is needed in order to minimise conflicts between resource-using activities, to enhance socio-economic opportunities (like optimising employment opportunities), and to bequeath an environmental estate for the benefit of future generations (Cloke and Park 1985). Hence the need for a local orientation of a sustainable development strategy. Quite often such a strategy is carried out by evaluating the implications of environmental standards or by putting constraints on industrial, land use, housing or transport developments.

In view of these observations, it seems meaningful, from a welfare viewpoint, to define local sustainable development in general terms now as *a development which ensures that the local population can attain and maintain an acceptable and non-declining level of welfare, without jeopardising the opportunities of people in adjacent areas*. This entails that local development is compatible with ecological conditions in the long run, while at the same time it does not interfere with supra-local (and preferably global) sustainable development (Van den Bergh 1991; Nijkamp *et al.* 1992). Consequently, a local sustainable development has to fulfil two goals: (1) it should ensure for the local population an acceptable level of welfare, which can be sustained in the future; and (2) it should not be in conflict with sustainable development at a supra-local level. These two objectives may be regarded in different perspectives leading to different interpretations of sustainability: weak versus strong sustainability, and internal versus overall sustainability. We proceed with our analysis by addressing these terms (see also Nijkamp and Opschoor 1997).

Internal versus overall sustainability

It seems plausible to argue that the biosphere provides mankind with a certain environmental utilization space (defined in terms of sustainable exploitation of renewable resources, land, absorptive capacities for pollution and waste, etc.). Within this environmental utilization space, subspaces may be defined in terms of the level of overall environmental pressure that these subspaces (nations, regions, localities) can sustainably exert (defined in terms of emission targets, nature conservation objectives, energy use reduction targets, etc.). From these national environmental utilisation spaces, urban economies may be regarded as having a fair share in proportion to e.g. their population size or their contribution to the national economy in terms of GDP.

A local treatment of sustainable development thus falls in between a macro (or global) systems level and a micro (or project) level approach, as the analysis of local or urban sustainable development is clearly a meso level analysis. Of course, local or urban sustainable development runs the danger of transferring the 'global rhetoric' (Pezzy 1989) of sustainability towards a micro or meso level of application. However, practical, efficiency or analytical reasons may necessitate a flexible use of different spatial scales. Also, management reasons may necessitate a differentiated level of treatment of local or urban sustainable development, both for legislative tasks and for control (or executive) tasks. Thus, local or urban sustainable development presupposes a meso level of analysis, though not on an a priori fixed spatial scale of a region or city.

Local or urban sustainable development differs from global or national sustainable development in three essential aspects: (i) the degree of openness (or interrelatedness) of a city or region; (ii) socio-economic imbalances which are, in contrast to global treatment, not levelled out on a local or regional scale; (iii) and the local or regional authority over common goods. These aspects should be considered explicitly to make sustainable development operational on the basis of (supra-)local, regional or supra-regional - but locally or regionally differentiated - policies. These three dimensions justify separate attention for local or urban sustainable development. Here we elaborate on these aspects by considering the scale at which urban development may generate environmental effects and what this may mean in terms of sustainability.

Urban economic activity makes use of environmental resources that are available in the cities considered: land, air, water, amenities, biodiversity. Implicitly it depends on the regenerative and absorptive processes underlying these resources; we refer to these processes as the 'environmental infrastructure'. Urban activities also draw on economic resources (produced capital, labour) and the social and economic infrastructure underlying them. In so far as these activities take place without adversely affecting the level of supply and the productivity of these resources and their infrastructural foundations (social, economic and environmental), one may say that these urban activities are 'internally sustainable', that is: sustainable with respect to the resource base of the urban agglomeration itself. However, both in terms of inputs into the urban economic activities and in terms of environmental outputs (pollution and waste), urban economies typically resort to environmental resources and environmental infrastructure beyond the internal urban resource base elsewhere in the country or even beyond national boundaries (cf. the above mentioned notion of 'an ecological footprint'). Energy and food resources or acid depositions are examples of this.

In a wider spatial sense, urban economies can be called sustainable if two conditions are met: (i) they should be internally sustainable as defined above, and (ii) their claim on resource bases elsewhere should not exceed their environmental utilisation space. If these two conditions are met, the urban economy is both internally and externally sustainable; i.e. it meets the conditions of what may be called 'overall sustainability'. Clearly, this condition will be hard to fulfil in practice.

Strong versus weak sustainability

When we take for granted that local sustainable development aims to attain economic and ecological co-evolutionary development at a local scale (Nijkamp 1991), we may interpret such a development as a non-negative change of local social welfare over time, with social welfare being made up of two correlated components (see also Nijkamp *et al.* 1992; Van Pelt *et al.* 1990). These are the consumption of man-made products and services (the socio-economic system) and the consumption of environmental amenities (the ecological system).

Clearly, the condition that social welfare changes should be positive may hold in two different ways, viz. either the change in one welfare constituent is positive and negative in the other one (provided the net overall change in social welfare is still positive) or the changes in both welfare constituents are non-negative. In the first case we speak of 'weak sustainability', whereas in the second case we face a case of 'strong sustainability' (Daly 1991; Opschoor 1992). Although these distinctions are conceptually sound, the absence of a 'numéraire' again makes an actual numerical illustration or application very difficult. Nevertheless, conceptually these terms can easily be handled.

It is clear that spatial substitution and substitution between welfare components are closely related phenomena. The above discussion can be summarised in the matrix of Table 1.1 on the scopes and domains at which sustainability may be defined, also in a local, urban or regional setting.

Table 1.1. Criteria for local sustainability

Domain / Scope	Internal sustainability	External sustainability	Overall sustainability
Weak sustainability	(w,i)-conditions	(w,e)-conditions	(w,e) + (w,i) conditions
Strong sustainability	(s,i)-conditions	(s,e)-conditions	(s,e) + (s,i) conditions

Source: Nijkamp and Opschoor (1997).

The (w,i)-conditions relate to the total local resource base (i.e. the sum of environmental and economic resources) and require non-negative changes over time. The (w,e)-conditions ensure that local claims on resources elsewhere remain within limits derived from sustainability standards at a higher (e.g. national) level; overall (weak) sustainability is achieved when both sets of criteria are met. Analogously, (s,i)-conditions relate to the economic and environmental resource bases (stocks and infrastructures) *separately* and require non-negative changes over time; (w,s)-conditions ensure that the urban claim on the natural (and economic) resource base remains within the environmental utilisation and economic possibilities frontier; overall (strong) sustainability implies that both sets of conditions must be met.

Seen from an urban sustainability perspective, we note that a city could, in principle, meet internal and external sustainability conditions by compensating (in environmental or economic terms) for the loss of specific resources or infrastructure, e.g. through environmental, economic and social development projects to be written off against current revenues drawn from the resource base. Thus, substitution between welfare components (especially in the weak sustainability approach), welfare and resource endowments over time and between spatial units could be envisaged.

In addition to the weak and strong variants of sustainability, one might also add a very strong variant of sustainability (Hunter 1998; Turner *et al.* 1994). In this approach, no utilisation would take place of non-renewable energies, because risk aversion in the avoidance of any potential damage to ecosystems could take precedence over the well-being of an impoverished human society. This stance, combined with a highly cautious view on technological innovation (which might ultimately improve the supply, treatment and use of energy) could enlarge the already high proportion of the global human population that has no access to an adequate and wholesome energy supply, both now and for future generations. From the viewpoint of intra- and inter-generation equity and the compliance with basic human needs, this degree of ecological determinism is hardly defensible. This very strong variant, which is a strict resource preservationist approach, is also unappealing when we look specifically at fossil energies as non-renewable natural resources.

After this extensive discussion of the spatial aspects of sustainability, we will in the next section devote attention to the potential role urban policy may play in achieving sustainable development objectives.

1.4 Sustainability and Cities

Sustainable development has of course a world-wide coverage, but there is a rapidly growing accordance that effective policy strategies also must have a meso orientation, e.g. by addressing specific economic sectors or regions. Therefore, a regional or urban angle for analysing sustainability in an operational sense is certainly warranted. Sustainability in an urban setting refers to the potential of a city to reach a level of socio-economic, demographic, environmental and technological performance which in the long run reinforces the foundations of the urban system itself. Thus, urban sustainability ensures a long-term continuity of the urban system under varying external conditions. It has to be added that in an open spatial system cross-boundary flows and external development stimuli may play an important role: insustainability may even be imported or exported (as witnessed in the notion of 'ecological footprint' alluded to above). In any case, the emphasis on the environmental potential of the city may enhance our insight into the feasibility of sustainability objectives formulated in a given institutional policy setting. In summary, sustainable cities are cities where socio-economic interests are brought together in harmony (co-evolution) with environmental and energy concerns in order to ensure continuity in change. A situation of non-sustainability of a city would imply a structural decline of the economic base of a city (reflected *inter alia* in population decline, environmental degradation, inefficient energy systems, loss of employment, emigration of industries and services, and unbalanced social-demographic composition). Environmental decay is usually one of the first signs of non-sustainability of an urban area (cf. Lynch 1981; Nijkamp and Opschoor 1997).

That the role of the locality in environmental management is increasingly coming to the fore, is witnessed by various policy initiatives. For example, the Commission of the European Communities (EC) launched its 'Greenbook on the Urban Environment' (1990), the Organization for Economic Cooperation and Development (OECD) published its report on *Environmental Policies for Cities in the 1990s* (OECD 1990), while many other institutions (international, national, regional and local) followed this new wave of interest in urban environmental research plans, by preparing urban quality of life programmes and the like. Various new concepts were advocated, such as the *green city*, the *eco-city*, the *eco-policy*, the *zero-emission city*, the *liveable city*, the *resourceful city*, or the *environmental city*. There is apparently nowadays a broad concern about the sustainable future of our cities. In the spirit of the Rio Conference, a broad array of local initiatives on environmental quality has emerged, often in the context of the above mentioned Local Agenda 21 (see Whittaker 1995). In Selman (1996) many issues centring around sustainable development from the viewpoint of opportunities offered by localities are discussed. This publication offers much information on the various roles of local authorities concerned with the daily quality of life, and addresses *inter alia* policy strategies, land-use planning aspects, corporate strategies for the environment, and the role of citizens.

Especially in the European context, the reinforced focus on the city seems warranted, as the European countries (especially those in the EU) are facing a stage of dramatic restructuring and transition as a consequence of the move towards the completion of the internal market. However, the aim to make Europe - and its cities - internationally more competitive in economic terms, may be at odds with its environmental sustainability. In the long history of Europe, numerous cities with an extremely valuable and vulnerable socio-cultural heritage have emerged which deserve strict protection in the interest of current and future generations. Therefore, what Europe is facing nowadays is a problem of ecologically sustainable urban development. Furthermore, European cities are faced with increasing traffic loads, while the goal of urban competitiveness is often at odds with urban sustainability.

Clearly, the urban interest in sustainable development is not only of a technical nature (e.g. by concentrating on district heating or solar energy), but concerns also policy coordination, land-use planning, education and training and demonstration projects. Are cities able to fulfil such roles? This question will be discussed in the next section.

1.5 Why Cities as Focal Points of Sustainability?

Sustainable city strategies have a 'Midas' character with many faces. We will now highlight various important advantages of the focus on energy policies in a sustainable city context. Urban areas are by definition centres of economic activity characterized by economies of scale and density. Given that a spatial clustering of activities implies a concentration of energy use, it seems plausible that urban areas are a suitable geographical entity as a focal point for energy policy (cf. Atkinson 1994). Admittedly, large energy consuming industries have usually relocated themselves from the core areas to the urban fringe, but that leaves the notion of urban areas as large concentrations of (direct and indirect) energy users, both for production and consumption, essentially unaffected (cf. Owens and Rickaby 1992; Owens 1996). It is certainly true that the majority of the world population lives in urban areas, and that makes these areas suitable as subjects of energy-environmental policies. Such areas also offer scale advantages in the production and distribution of energy (see Camagni *et al.* 1998).

In the light of the above observations, one may list several reasons why a well focused energy-environmental planning strategy at the urban level is a promising activity in the framework of sustainable urban development and why there are good reasons to concentrate on the urban dimension of energy policy.

The major advantages for addressing local policy-makers in coping with global energy and environmental issues are (see Nijkamp and Ursem 1998a):

- a significant share of the world population - and industrial activities - is found in urban areas, so that it makes sense to concentrate efficient energy initiatives on these areas in order to benefit from agglomeration economies in the energy policy sector

- at a local level a wealth of practical insights and ideas is available on how to cope with inefficiencies in energy supply, energy use and environmental recycling potentials (e.g. biomass)
- the current decentralisation movement opens up many possibilities for local authorities to be actively involved in building up effective and operational energy savings plans and related environmental quality improvement initiatives
- urban areas are becoming recognized institutional policy units with a clear competence and with the possibility to operate in a flexible and innovative manner
- an urban orientation of energy and environmental policy may also encourage a direct involvement of citizens, as such policy initiatives are usually source-based, effect-oriented and visible, so that a sufficient local support base may be generated
- finally, urban areas are also a suitable spatial scale for systematic data collection, monitoring and analysis of proper energy/environmental indicators.

In short, cities certainly qualify as focal points for sustainability research and planning, as they play a decisive role as nodal points of people and their activities. In many cases they also face the most severe environmental problems, such as air and water pollution, noise, annoyance, waste, declining quality of life, and destruction of urban landscapes and architecture. This also explains the current heightened public awareness and concern about the quality of the urban environment (including also public health). The broad nature of urban sustainability suggests also that urban policies aiming to achieve sustainable development should be strategic in nature, integrative, and visionary regarding the role of the private sector, focused on the provision of market incentives, and more oriented towards the needs of citizens (see Davidson 1988, Selman 1996, Stren *et al.* 1995, White 1994).

1.6 Strategies for Urban Sustainable Development

An environmentally sustainable development of a city can only be attained by initiating appropriate policy strategies. On this subject much literature can be found which focuses on the design of concepts or frameworks needed for such policies. It is clear that initiatives in various cities world-wide differ strongly in the adoption and implementation of such concepts, because each city has its own specific geographical, political and environmental setting. Nevertheless, general integrative concepts and evaluation procedures may be developed which can serve as guidelines for many cities undertaking sustainability initiatives. A broad survey of such concepts can be found in Selman (1996), while an overview of policy strategies can be found in OECD (1995). Although it is likely that environmental quality problems may become more severe with urban size, there is no clear evidence that urban size as such causes environmental decay. According to Orishimo (1982) it is not the sheer city size, but rather the implied land use, the transport system and the spatial layout of a city which are critical factors for urban environmental quality (see also Nijkamp 1996).

Policies addressing sustainable development of cities should, therefore, cover multiple fields, like urban rehabilitation, urban land use, urban transport systems, urban energy management, urban architecture and conservation policy, and urban cultural policy. Measurable indicators including minimum performance levels and critical threshold levels will then have to be defined, estimated and used as forecasting tools so as to improve awareness of sustainable development issues of modern cities. Local authorities will have to share their tasks with all other actors in the urban space (including the private sector) in enforcing and maintaining these critical thresholds. It goes without saying that urban sustainable development is a process rife with conflicts and incompatibilities. Commitment to a strict environmentally sustainable urban development by key actors in a city is necessary for a successful implementation of sustainability policies. In doing so also economic (market-based) incentives are desirable in order to increase efficiency and to cope with the negative factors of modern city life. Failure to develop an effective balanced urban development policy will reinforce urban sprawl and will highlight inner city problems to a much larger area. Environmental-benign urban policies may, on the other hand, attract new investments, favour urban employment, and hence contribute to an increase in quality of life. In general, three major fields of policy interventions may be distinguished:

- a reduction of the (growth in the) actual use of scarce energy (e.g. through behavioural change)
- an increase in technical energy efficiency (e.g. insulation, heat pumps, district heating)
- a wide-spread introduction of new sustainable energy technologies (e.g. photovoltaic systems, wind energy).

The successfulness of such interventions depends clearly on three major background determinants:

- *institutional factors* (management and organisation of the urban energy sector, public-private modes of cooperation etc.)
- *attitudes and behaviour of citizens* (life styles, mobility patterns, environmental awareness etc.) (cf. Van Poll 1997)
- *urban structure and morphology* (population density, urban form, transportation network etc.) (cf. Breheny 1994; Jenks *et al.* 1996).

Local authorities have the possibility to exert both a direct and indirect influence on these determinants.

The question whether a given urban development is sustainable or not is co-determined by the targets set by policy-makers. There is not a single unambiguous urban sustainability measure, but a multitude of quantifiable criteria which may be used in an empirical test. A necessary condition for implementing an effective planning system for urban environmental management geared towards maintaining sustainability is the development of a system of suitable *urban environmental indicators* (see OECD 1978). Such indicators, which should represent a balance between the necessary quality of information and the costs involved, would have to be related to economic, social, spatial and cultural dimensions of the city. The OECD has drawn up a long list of elements which are decisive for urban environmental quality and which would have to be included in such an indicator system. Examples are: housing, services and employment, ambient environment and nuisances, social and cultural concerns, etc. However, it appears to be extremely difficult to operationalize such an indicator system. This means that precise empirical evidence on urban environmental quality and on the implications for both household and firm behaviour is not always available.

In light of the previous observations, the conclusion seems warranted that the road towards sustainable cities is not an easy one. In the sequel of this book we will show that interesting and successful examples of strategic and integrative approaches to urban sustainability do exist; lessons should be drawn from them, and extended to other cases. This also implies that a meta-analysis of environmental policies in the city may be meaningful, as it may generate transferable lessons.

1.7 Urban Sustainable Development and Energy Use

The previous section has demonstrated some cornerstones of sustainable city policy. In order to safeguard the socio-economic and environmental potential of modern cities needed to host the future generation, local energy and environmental planning is increasingly recognized to be a *sine qua non* condition. Sustainable cities aim at achieving a balanced (co-evolutionary) development in which economic forces (e.g. efficiency), social considerations (e.g. equity and access to facilities) and environmental concerns (e.g. quality of life) are brought together from the viewpoint of a 'green society' (see also Pearce *et al.* 1989). Enhancing local energy efficiency is likely to be one of the critical success factors for sustainable cities. The basic thought behind this is that a higher energy efficiency cuts with two edges; namely it reduces the actual depletion of natural energy resource stocks, while it also reduces the environmental pollution caused by the conversion to usable energy. This 'double dividend' effect is even more important, considering that usually energy has to be imported to the city, placing difficulties to the city to reach an external (and thus overall) sustainability regarding energy use. Another important reason for local energy efficiency for the growth of a sustainable city is that energy management provides substantial support to the improvement of the local economy (i.e. higher degree of competitiveness), and to a more affordable - and hence equitable - distribution of scarce resources, as well as a reduction in the environmental burden.

Innovative energy programmes may contribute to urban sustainability by a better reservation of the environmental resource base. In the mean time, they lead to an important enlargement of the environmental infrastructure, which is in fact the other component of the resource base that should offer future generations equal access to possibilities to use the natural environment as an input into the process of their welfare generation. Urban environmental quality requires an active urban energy policy reducing the consumption of fossil fuels. By strict energy savings measures a considerable part of pollution caused by households, industries and transportation can be avoided.

While a favourable environmental quality in cities is necessary for a balanced urban economic development, urban development is not necessarily at odds with a favourable environmental quality, and certainly not in the long run. The previous scheme is based on a balance between urban development, environmental quality and energy efficiency; these different forces are also schematized in Figure 1.1.

An important issue is the financial feasibility of urban energy programmes within the context of urban sustainability. Even when such programmes are physically feasible, this does not imply that they will actually materialize. Possibilities for financing are one of the most critical conditions for a substantial improvement of urban quality-of-life conditions. Does an urban energy policy within the context of sustainability generate sufficient 'value for money' which would justify the mobilisation of huge financial resources? Similar questions have recently also been raised in the context of other 'non-productive' assets such as housing, infrastructure and culture (cf. Struyk *et al.* 1990). Despite the indigenous merits of housing, infrastructure and culture, it was necessary to also show that these assets would create a positive contribution to social welfare to such an extent that it was worth investing in them. In effect, what is required is that a net positive contribution is made to the present value of flows of incremental social welfare over time. But this net present value is determined to a large degree by time preferences and hence by discount rates. Especially in the context of developing countries this means that the net contributions of sustainability-oriented urban programmes may appear as relatively low, given the relatively high prevailing rates of individual and enterpreneurial time preference.

It is evident that net positive socio-economic benefits are a *sine qua non* to justify public intervention in the urban environment of energy supply.

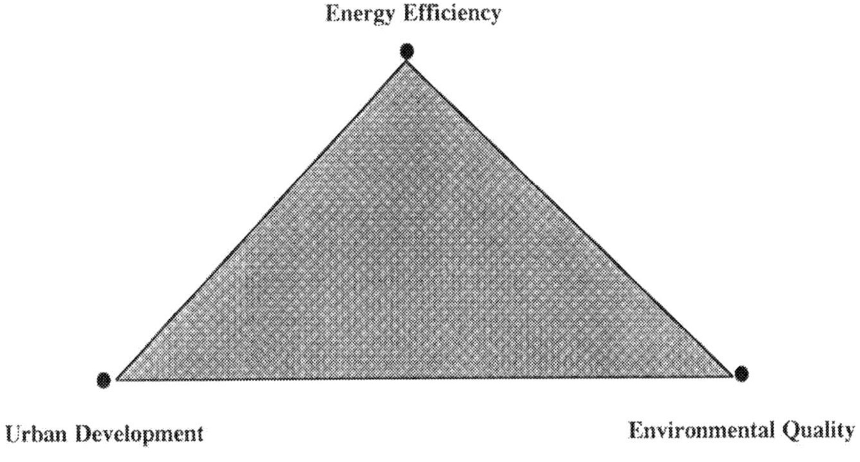

Figure 1.1. A triangular force field of urban growth, environmental quality and energy efficience

1.8 Energy Policies for Urban Sustainability

There is not a single strategy for enhancing the environmental and energy sustainability of cities. In practice, a variety of options exist for a sustainable urban energy policy. Household activities and consumption, industrial and commercial activities, and transportation are - in addition to electricity production - the main sources of fossil energy use and air pollution. Many European experiences have shown that considerable savings in fossil energy use and reductions in air pollution are still possible in all these sectors. Roughly spoken, there are five possible policy fields by which these improvements can be realized. These may concisely be listed as follows:

- urban land-use planning and urban architecture;
- conservation and strict regulatory energy policies;
- stimulation programmes (e.g. financial incentives for industrial activities and information campaigns for the public);
- market-based energy policies;
- policies encouraging technological innovations.

Up till now, most of the research on and experience with urban sustainability policies has focused on the first three issues (see, amongst others, Selman 1996). Policies that have been less applied are market-based energy policies (except in the USA). Chapter 2 of this book is devoted to a further exploration of physical planning (or land-use planning) options and market-based approaches. However, the main focus of this book is on those options that in recent years have shown a rapidly growing and promising potential, namely policies based on technological innovations in the energy sector. We will give here a brief description of the state-of-the-art in this area.

In *industry* new technologies and better insulation of buildings may lead to a considerable rise in energy efficiency, although this clearly has a long lead time in normal circumstances. In the *residential* sector, housing insulation programmes may also lead to drastic energy savings for both space heating and air conditioning (e.g. by means of better insulation, heat pumps, solar energy installations, wind turbines, and economizers for central heating systems). Also in the *transport* sector considerable savings are in principle possible (e.g. through more energy-efficient engines, vehicle weight reduction or - in the long run - through a more energy-efficient physical planning aimed at a reduction of commuting distance and/or a shift of the modal split in favour of public transport) (see also Button 1992). At a more integrated and meso level of urban energy planning, various possibilities are offered by central heat distribution, recycling of energy from heat, by combined heat and power either in district heating or in co-generation, or by using urban/industrial waste as a fuel for generating plants. Especially at a local level these energy saving options are likely to be more efficient than at a more region-wide level, as in general such options require fairly high densities of energy demand. In addition, land use zoning and physical planning might also provide sufficient scope for necessary regulatory regimes for environmental policy.

In general, urban energy planning based on technological innovations may comprise a whole set of different and complementary energy policy strategies ('packaging' of policy measures), such as industrial cogeneration, district heating, combined heat and power (CHP) generation (using steam turbines, internal combustion engines, gas turbines or combined cycle gas turbines), PV systems, combined urban waste management and energy production, load management, and institutional reforms in the structure of utilities. Although most of the ideas in the present publication hold for the entire energy sector, we have in our empirical work addressed, in particular, renewable urban energy systems, as these systems are as yet under-researched and too often neglected.

In all cases, environmental programmes are expensive and require much public money. In light of severe public budget constraints and deficits of governments at all levels (national, regional, local), a strong plea seems in place for private financing of public programmes or projects, especially in the wake of the deregulation movement. In most cases however, fully private financing is not regarded as a desirable option. Hence in practice, we see the emergence of various types of public private partnership (PPP) configurations. Our analysis earlier in this section suggests the need to engage in such partnerships, especially when public and private rates of time preference or public and private valuation for environmental quality diverge.

Defining and implementing a sustainable urban energy policy is a formidable task which urgently needs political attention. Such policy can be defined with various domains and scopes in mind; the more stringent one chooses to be, the lower the physically feasible sustainable level of socio-economic welfare. The less stringent one is, the more the costs of actual urban development are shifted on to the city's own future population, or on to other parts of the economy beyond the urban sphere, either now or in future. Urban sustainability in the supply and consumption of energy is a challenge for creative urban policy and for technological and administrative innovators. Prime movers in the direction of innovative energy programmes may find this to be beneficial, not only in environmental terms but also in socio-economic terms. At the same time innovative energy initiatives pose complicated questions regarding private-public cooperation and financial matching. Some of these institutional issues will also be discussed in Chapter 2.

2 Institutional Regimes for Sustainable Urban Development

2.1 Urban Policy Opportunities and Impediments

Sustainable development is not 'manna from heaven' which will automatically come into being. It requires active involvement of all actors and stakeholders in both the public and private domain. Cities are important players in this perspective, not only in an environmental context (as highlighted in Chapter 1), but also in an energy context. The increasing returns in urban agglomerations offer many opportunities for efficiency gains and economies of density in the energy sector (see Gilbert *et al.* 1996):

- relatively low per capita costs for the construction and maintenance of energy infrastructure provisions
- a sufficiently large critical mass of people and activities for an efficient reduction of energy consumption
- a sufficient density of people and workplaces to stimulate energy-efficient forms of public transport
- a seedbed for the development and implementation of creative forms of new environmental and energy strategies and policy approaches.

The opportunities for efficient energy policies in urban areas will certainly rise in the light of the increasing world-wide urbanisation rate. Consequently, local authorities may play an important role as coordinator and animator of an integrated energy-environmental policy, which is characterized by flexibility, innovation and public support.

Clearly, there are also severe bottlenecks to be envisaged in the implementation of urban energy policy tasks. For example, the Urban Energy Handbook of the OECD (1995) states: "*The vast majority of municipalities lack the up-front capital to implement energy-efficient improvements on their own*". Apart from limitations imposed by scarce financial means, there may also be severe impediments caused by the lack of necessary cooperation between network owners and small-scale suppliers of energy. Especially institutional inertia regarding urban partnership (e.g. flexible third party access) appears to be a major stumbling-block for efficient and effective decentralized energy initiatives (Nijkamp and Perrels 1994). Seen from this perspective, one may plausibly argue that learning mechanisms among local or regional actors sharing common concerns may lead to competitive advantages for urban areas.

In general, there appears to be a serious lack of market orientation in the urban energy sector. Nevertheless, cities may be seen as actors *par excellence* to encourage both the introduction of renewable energy forms and the savings of non-renewable energy consumption (see also Camagni *et al.* 1997). Such a policy will not only reduce location costs of business and residential activities in urban areas, but will also improve the quality of life in urban areas, a factor which is increasingly regarded as a major location factor in a modern society (see Van Geenhuizen and Nijkamp 1997).

In the light of the above observations, good governance supported by all interest parties is a *sine qua non* for successful urban sustainability policy. There are however, different types of policy-making regimes and policy orientations. In Part B of the present book we will extensively focus on innovation policy. In the present chapter we will mainly address institutional (regulatory and market-based) approaches.

An intriguing question is whether - without leaning on technological innovations - effective governance strategies in urban energy production and use can be developed that contribute to sustainable urban development, as described in the previous chapter (see also Camagni *et al.* 1997). In the context of sustainable urban development, it ought to be recognized that a proper organization and use of the urban environmental space and energy resources should exploit scale and urbanisation advantages. Due attention has to be paid to a balanced choice out of the portfolio of innovative capabilities of the city (and of firms located in the city). While for this aim dedicated policies focused on urban sustainability are needed, thus far the results of urban environmental policies have not been overwhelmingly impressive, despite a multiplicity of regulatory systems.

Given the fact that a large share of air pollution in cities is related to the production of energy and the use of fossil energy resources, and that the use of energy implies an important depletion of the environmental resource base, it is obvious that any governance strategy should focus on these aspects. This chapter explores the potential of two regulatory institutional options, viz. physical planning (in particular, compact city plans) and market-based policy strategies. We will argue that systems of tradable permits for air pollution and energy consumption may be new promising instruments to ensure urban sustainability through favouring efficiency, equity and environmental quality in the city. An overview of such market-based policies will be given, and their principles (which are more general than those of just the energy sector) will concisely be set out. The focus will however, remain on urban energy production and energy use.

2.2 Physical Planning, the Compact City and Environmental Space Conflicts

The city is a compact spatial configuration of economic activities. Through its compact form various agglomeration economies can be realized. The spatial distribution of activities in the city does not guarantee however, that the most efficient lay-out from an environmental or energy perspective is attained. Through a relocation of activities more energy-efficient or environmentally-benign patterns might be realized. However, in general, this is a complicated issue, as spatial accessibility, land use prices and distribution of different income groups are closely connected phenomena, so that interventions in the one field will have impacts on other components of the urban system. This also applies to compact city initiatives.

In recent years the concept of the compact city solution has strongly been advocated to solve the urban sustainability problem (EC 1990; Elkin *et al.* 1991). Consequently, many urban governments nowadays try to apply a compact city solution in their strategic urban plans. A compact city is a city which has a high density and a concentration of socio-economic functions in order to reduce energy use, environmental decay and urban sprawl.

There are indeed many advantages involved in high urban density living (see Haughton and Hunter 1994). High density living is more efficient than low density modes of living in terms of lower rates of domestic consumption of energy for heating, electricity and water. Scale economies can result in lower infrastructure costs,

and critical mass thresholds are more likely to be reached to justify the provision of energy-efficient modes of public transport (underground metros, trams, etc.). In addition, high density urban development is regarded to economize on street lighting and to reduce transport needs, particularly where cities are arranged in such a way that people can readily walk or cycle to their work and services, such as schools and shops. A further positive feature concerns the range of functions which well-designed neighbourhoods can support with high concentration of population, in terms of health care, schools, shops and so on. Newman and Kenworthy (1989) have even found a strong correlation between the density of the urban area and the energy use in it. The higher the density of the city, the less energy is used. Some other authors however argue that, although there is a strong correlation, also many other factors are involved like the size and the urban form (Breheny 1993).
The advantages of concentrating urban functions in a compact city may be summarized as follows (Rosdorff *et al.* 1994):

- by concentrating economic functions in the city, the entire area affected by urban pollution or environmental decay, is reduced;
- a reduction of environmental stress per activity due to scale economies as a result of improved efficiency in both the economy and the resource use.

However, the compact city has also a major disadvantage. By a geographical concentration of the functions of the city, also the spatial concentration of environmental pollution is higher: because environmental stress is spatially more concentrated, the environmental problems are more severe as well in a smaller high density area. This influences the quality of life of people in the compact city. This situation is also called the *paradox of the compact city* (Breheny 1993). We may argue that this paradox is an allocation problem of environmental assets (clean air, clean water, etc.) to space and time (Rosdorff *et al.* 1994). In the compact city the environmental assets are overused in order to reduce the use of environmental assets outside this area.

In the light of the tension between economic efficiency, social equity and environmental sustainability in a compact city model, it is conceivable that many policy initiatives have been launched which were seeking for an improvement of urban quality of life through more strict regulation (e.g. environmental standards etc.) In many countries the government works predominantly with norms for environmental pollution (e.g. dBA for noise, mg/m^3 for air pollution). In terms of strict efficiency, local standards would appear preferable to the adoption of nation-wide emission standards. The latter do not guarantee that the marginal costs of environmental improvement will be equalized across urban areas and, in consequence, the uniform standard will be too generous for some cities and too severe for others (Button and Pearce 1989; Orishi 1994).

With the present legislation, a company in an urban area is not allowed to produce a level of pollution that exceeds the standard set by the government. The norms however, are often rather uniform and may be more or less equal for e.g. an industrial area or for a natural park. This means that sometimes it is not possible to locate industrial firms in close proximity, because the norm of the area is exceeded. The result is that in some areas the development is halted, because it is not possible for new firms to locate there, whereas in 'clean' areas the development will likely start to grow. When a new industrial firm is not allowed to locate near an existing industrial area, there is the danger that the pollution gets even more dispersed. Given the present legislation in many countries, it may thus be difficult to implement a compact city with a balance between industrial and residential (cf. Yanarella and Levine 1993).

Another disadvantage of the present legislation is that there are no incentives for companies to reduce their pollution level below the norm, even if the company is able to do so at low costs. Consequently, it seems meaningful to look for a policy which is more flexible and offers firms more incentives to reduce their pollution. Flexibility is then needed to concentrate the pollution problems and efficiency is needed to reduce these environmental problems at the lowest costs possible. In the USA, several initiatives have been started in the past decade to find a solution to these problems, mainly based on market principles with a particular emphasis on the idea of emission trading. This system will concisely be described in the next section.

2.3 A System of Marketable Emissions Permits

Emission trading is a relatively new instrument in environmental policy (see Klaassen 1996; Koutstaal 1997). In a marketable emission permit system (sometimes also denoted as tradable emission permit system or, in short, permit system) the government or an environmental authority allocates or sells emission permits to companies, which can trade their emissions afterwards. Some companies will be able to reduce their emissions (or the impact of the emissions) at relatively low costs (low marginal abatement costs). They can then trade the emission permits which they do not use to companies which are not able to reduce their emissions due to relatively high marginal abatement costs. In a perfect market the emission permits will be traded until the marginal abatement costs of the companies are equal to the price of the emission permits. This means that the reduction of the emissions are allocated in such a way that the reduction is realized at the lowest costs (Montgomery 1972). Clearly, this outcome is only reached in a perfect market. The requirements for a competitive tradable permit market are (Hahn 1984):

- the existence of a large number of potential traders;
- an 'arm length' (non-dependent) relationship among polluters in their product markets to prevent subversion of permit trading through product market threats;
- sufficient differences in abatement costs among the potential traders to make trading worthwhile.

There are broadly two forms of permit markets (see also Nijkamp and Ursem 1998a). In the *Ambient Permit System* (APS) (also called the Ambient Differentiated Permit (ADP)) the permits do not concern a source emission, but refer to the impact of these emissions at the level of pollution at a particular point (a receptor point). Environmental quality norms are made for each receptor point and competitive bidding for these permits per receptor points would then generate an equilibrium solution that satisfies the conditions for the minimization of total abatement costs (Montgomery 1972). A main advantage of this system is the simplicity of the system for the environmental authority. In particular, officials need no information or whatsoever regarding abatement costs; they simply issue the prescribed number of permits at each receptor point, and competitive bidding takes care of matters from there. But a great disadvantage is that it is extremely cumbersome for polluters. The polluters have to make a portfolio of permits from each of the receptor points that is affected by their emissions. When there is a large number of receptor points, each with its specific environmental norm, a huge number of markets will exist. The transaction costs for polluters may then be high (Baumol and Oates 1988). A second disadvantage of the APS is that the receptor points tend to become institutionalized. They all have their specific markets and the displacement of a receptor point to adapt it to a new pattern of pollution would create dislocations. The APS form of the permit market is thus not without serious problems (Baumol and Oates 1988).

The second form of permit markets is the *Emission Permit System* (EPS) (also called the Emission Discharge Permit (EDP)). The EPS can simplify life significantly for polluters. Instead of assembling the necessary portfolio of permits from different receptor markets, each source would find itself assigned to a single zone within which emissions entitlements would exchange one-for-one, but no transfers are permitted between the zones. However, the EPS cannot, in general, achieve the least-cost solution, and it places high demands on an administering agency that aims to approximate this solution. Since polluters with somewhat varying dispersion coefficients are aggregated into the same zone, one for one trades of pollution entitlements will ignore the differences in the concentrations contributed by their respective emissions. In short, the price of emissions for each polluter will not correspond accurately to the shadow price of the binding pollution constraint (Tietenberg 1980).

Since both the APS and EPS forms of marketable permit systems are subject to some serious problems, a number of hybrid systems has emerged. One of these hybrid systems is the *pollution offset-system* (PO).

The emission offset policy was originally designed as a means for allowing economic growth in non-attainment areas while insuring no further degradation of their air quality. In this approach, permits are defined in terms of emissions (e.g. the permit allows the discharge of X tonnes of the pollutant, say, per month). However, sources are not allowed to trade permits on a one-to-one basis. More specifically, transfers of permits under the PO scheme are subject to the restriction that the transfer does not result in a violation of the environmental quality standard at any receptor point. The new source must, for instance, always use the 'lowest achievable emission rate' technology.

The advantages of this hybrid system are significant. The PO system shares with the APS the important property that mutually beneficial trades among sources can lead to the least cost solution, a result which is independent of the initial allocation of the permits. The PO scheme makes modest information demands on the environmental authority. Such authorities need to know the dispersion characteristics of emissions within the zone, but need not have any information on source abatement costs. Thus, the authority does not have to solve the cost minimization problem to determine the initial allocation of permits; any allocation will do.

Unlike the APS, the PO system does not require sources to trade in a multitude of separate permit markets. Instead, a firm purchases emission permits directly from other sources. The PO scheme thus promises substantial savings in transaction costs to sources relative to APS. In addition, it is not subject (as is APS) to the requirement that a fixed and 'institutionalized' set of receptor points be established (Baumol and Oates 1988).

Another related concept which has emerged is the *bubble concept*. The bubble concept specifically allows emitters to propose modifications in their emission standards, based on the substitution of a more stringent degree of control to another source of the same pollutant. The attention is not focused on one source only, but on a group of sources. The pollution by the group of sources - rather than a single source - should not exceed the total norm. In the bubble concept a bubble is put over the group of sources and within the bubble the companies are free in how far they want to realize the norm for the bubble. These substitutions can, under certain circumstances, take place between plants or even between firms (the possibility of compensating sources within the own company is also a form of a bubble and is often called *netting*). This design feature carries the bubble policy a long way toward a fully transferable permit system. The aim, of course, is to allow a firm to meet its emission reduction goal as flexible and cheaply as possible while insuring that the air quality is not degraded by the substitution (Tietenberg 1980).

The main difference between the bubble and the offset policy is that the latter allows the transfer of emission reductions from existing sources to new sources, as long as there is a net improvement in the environmental quality, whereas the bubble policy allows transfers only among existing emitters and does not ask for a net improvement of environmental quality.

The key difference between the existing system (with the bubble concept and the pollution offset system) and a system with full transferability lies in the fact that in the latter the control authority allows all sources to participate in the trades and allows all emission reductions to be traded in a regulated market. In contrast, the bubble and offset policies have restrictions on which emission reductions can be traded (e.g. only those additional reductions above the standard in the offset policy) and on which sources can participate in trades (e.g. only existing sources in areas demonstrating attainment in the bubble policy) (Tietenberg 1980).

Now the interesting question is whether a policy of marketable emissions is consistent with the concept of urban sustainable development. The policy of marketable emissions sets standards for the environmental quality in a specific area, and these standards may be high when the interest in urban environmental quality is high. It is thus in principle possible to reach urban sustainable development with modified economic growth through the policy of marketable permits. This will be further discussed in the next section.

2.4 Marketable Emissions and Urban Sustainable Development

An important advantage of the marketable emissions concept is that we only have to know to which degree we want to pollute or use scarce natural resources. We simply set standards at these points without the difficult task of valuing the environment. With other policies (like the 'polluter pays' policy) we have to know the value in order to calculate the tax for the firm. Thus the simplicity of the policy of marketable emissions makes it an attractive policy (Howe 1994).

Furthermore, it is noteworthy that sustainable development is an integrative process. The policy of marketable emissions tries to integrate environmental concerns into the economic system: the environment is seen as a scarce resource. In Section 2.2 the compact city was mentioned as a solution to the problem of urban sustainable development. Theoretically a policy of tradable emissions leads to a state where the environmental goals are reached at the lowest costs (see Montgomery 1972; Baumol and Oates 1988), so that the system of tradable emissions is efficient in reducing urban environmental problems. Clearly, this policy is not the only policy that achieves environmental goals at lowest costs. The system of effluent fees also has this advantage. But a major advantage of marketable permits over effluent fees is that permits promise to reduce the uncertainty and adjustment costs involved in attaining legally required levels of environmental quality. The environmental authority is normally never entirely certain of the response of polluters to a particular level of effluent charges The fee may have to be raised and then altered again to generate an iterative path converging toward the target level of emissions. In contrast, under a permit scheme, the environmental authority directly sets the total amount of emissions at the allowable standard; there is, in principle, no problem in achieving the target (Baumol and Oates 1988).

Next, a tradable emission policy is flexible, as it allows the environmental agency to set different standards for different geographical locations. In this way it is possible to have areas which demand more attention for the natural environment, and other areas which call for more attention to economic growth. With the system of emission trading, it is also possible for a firm to expand or relocate to areas where this was previously not possible under the present legislation. The system has the possibility of financial substitution for emission reduction. With the possibility to buy, sell or bank the emission reduction credits, there are many incentives for firms to reduce their emissions. Like all policies, the concept of marketable emissions also embodies a couple of disadvantages. In the first place, storing emission reduction credits at a bank creates a right to pollute. The idea that a company can win the right to pollute and therefore harm third parties with its pollution, is somewhat counter-intuitive. But there is an important distinction between 'emission reductions' (the *physical* reduction of emissions by a source) and Emission Reduction Credits (the *commodity* that is to be banked). The simple fact that a source reduces its emissions by 100 tons, does not necessarily mean that either the source itself or another source that purchases these credits has the right to pollute an additional 100 tons. The use of Emission Reduction Credits is subject to the rules governing the particular permit context (e.g. offsets or bubble applications) to which it is applied. In order to avoid confusion between the physical pollution units and the intangible commodity, which is banked and ultimately used, the former ones are termed emission reduction and the latter Emission Reduction Credits (ERCs). Credits will entitle sources to meet certain requirements for obtaining new source permits or for meeting emission limits at existing sources. The use of ERCs must stand up to sufficiency and equivalent tests. Credits do not automatically cntitle sources to increase their pollution; nor are these credits absolute entitlements or property rights of interminable duration.

A second disadvantage of the marketable emission system is that it is not entirely fair to newcomers in an area, who have to pay for permits, while the already existing companies do not have to pay for them. This is only the case when the initial distribution is done by grandfathering. Grandfathering means that the sources which are in the area, are given the permits at the initial distribution (usually on the basis of historical emissions). Trade between the holders will result in an equilibrium solution that satisfies the conditions for the minimization of total abatement costs. Another system that could be applied, is that the permits are sold at the initial distribution. Auctioning the permits results in an efficient initial allocation. In this latter system, the costs for the initial allocation may be high for the sources in the area. Grandfathering therefore has the advantage that it is far more politically feasible. In all cases of tradable emission permits, the initial distribution was done through grandfathering. Another possibility is that the authority reserves a share of permits for possible newcomers in the area. This is being done in a Swedish study, in which 80% of the permits are distributed through grandfathering and 20% are sold. The permits that are distributed through grandfathering have to pay a yearly tax on the permits they own. But this system of paying taxes eventually is the same as when the permits are sold initially (MEZ 1992).

A final, often mentioned, disadvantage against the bubble policy is the danger of filling up the norms. Within the bubble, companies are free to choose how they achieve norms. Consequently, the norms are totally used up, because a reduction credit may be sold or used for this purpose. But it is questionable whether this is really a disadvantage of the system. The norms are related to the number of permits that exist in an area. If the norms are set high enough, and are adjusted in the future, the problem is solved.

As was mentioned above, in order to offset, the total environmental quality must be improved. It is thus not entirely true that the emission reduction credits may be used to fill up norms.

After this comprehensive introduction to marketable emission permits, we will in the next section discuss their relevance for urban energy policy.

2.5 Tradeable Permits and Energy Use

2.5.1 Introduction

The possibility of implementing a system of tradable emissions for different environmental problems in the urban area would administratively be easier, if all types of pollutants could be handled within the context of one general market (cf. Verhoef *et al.* 1997). The problem to this approach is, however, that the final amount of each pollutant is determined by the emitters so that the control authorities control of individual pollutants (Tietenberg 1980). Therefore, it is better to use separate markets for separate pollutants, provided the number of different markets is not excessive. With many different markets, the administration costs will be high and subsequently also the transaction costs for polluters. Therefore, it seems more plausible to distinguish different markets for different pollutants in the urban area. The first step in designing a permit system is to define a basis for what is to be traded. Theoretically, there are three ways of managing environmental problems:

- control of the input that causes the pollution;
- control of the emissions of the pollution;
- control of the impact of the pollution on the environment.

We have to keep in mind the following criteria for the basis of the permit (MEZ 1992):

- the system should not be too difficult to implement; important aspects are the information required and the costs of control.
- the policy instrument must be effective, which means that there must be a direct link to the goal of pollution reduction.
- the system must have the possibility to control different pollutants within the same market; which goes to say that the number of different markets should be limited.

- the system should be dynamically efficient whereby developments of new technologies to reduce pollution are stimulated through the use of the permit.

We will illustrate the above observations by taking a closer look at urban air pollution. From the viewpoint of minimizing administration and transaction costs, it would be best for all different air pollutants to be traded in the same market. Macintosh (1973) suggests choosing the weights for determining the equivalency between different pollutants on the basis of their historical contribution to pollution in the local area. The permits in his study cover particles, sulphur oxides and carbon monoxide. Macintosh uses the emission of the different air pollutants as the basis for the permit. The disadvantage is that the weights that have to be chosen are difficult to establish, and may change over time. A control on each pollutant will be lost, since substitution between the pollutants is possible. A second disadvantage is that some air pollutants have a more local character than others. This could cause certain hot spots in some areas.

When we take a look at the sources of pollution, we can see from Table 2.1 that most air pollutants are, in one way or another, related to the use of fossil fuels for energy. A trading system based on the use of fossil fuels might therefore also reduce the concentration of different air pollutants. When the fossil fuels are completely combusted, there is a direct relation to the emission of CO_2 in the air. With an incomplete combustion, the emission of CO_2 is less, but the emission of CO and VOC's is higher.

One advantage of using fossil fuels as a base for the urban permit system is that the system is far easier to control than a system in which the emission of pollution is controlled, since registering the emission from all sources is a difficult task. The big advantage in choosing carbon based fuels is that data on production, import and consumption of carbon based fuels are usually available from the energy company or the gas company (MEZ 1992). A second advantage of using fossil fuels as the base is that we do not have to use different markets for all air pollutants. There is also a rather straight- forward connection to the amount of air pollution, while the system is rather flexible.

The second step in designing an urban permit market is to distribute the permits to the actors in the market. In the energy market, we can basically identify three groups of consumers:

- the industrial sector;
- the residential and commercial sector;
- the transport sector.

A trading system with one market for the different consumers of energy is difficult to design, since the difference in consumption of energy is significant. In the residential sector, the number of participants is large, but the amount they use is relatively low. In the industrial sector the number of consumers is relatively low, but the consumption on the other hand is high. A trading system in which all actors participate is thus cumbersome, because of the different sizes of the permits. For the practicability of the system of tradable emissions, the costs of control and the required information should not be too high, as otherwise the administration and transaction costs will be too high. Therefore it seems better to use separate markets for different consumers of energy.

Table 2.1. Sources of air pollution

Air pollutants	Source
Carbon dioxide	Combustion of fossil fuels
Nitrogen oxide	Transport sector, industrial sector, energy sector
Carbon monoxide	Incomplete combustion of fossil fuels
Volatile organic compounds (VOC)	Incomplete combustion of fossil fuels
Particulate matter	Sources include fine asbestos and other particles stemming from wear and tear of tires and brakes as well as matter resulting from engine, especially diesel engine, combustion
Sulphur dioxide	Coal fired electricity generation, and for a small part the transport sector (5%)

2.5.2 The industrial sector

One of the main energy consumers in the urban area is the industrial sector. The share of the total amount of energy consumed by industries in OECD countries, is about one third (OECD 1995). This share has declined in the past 25 years (from 40.1% in 1970 to 32.1% in 1991). This is partly due to a more efficient energy use, but also because of a structural shift in the economic base of most industrialised countries to a more service-oriented economy. The environmental problems directly caused by the use of energy by the industry are the depletion of energy resources and air pollution. The share of the different emissions of gases is shown in Table 2.2.

A permit system for energy used by the industrial sector will reduce the gas emission released by the combustion of fossil fuels. The cost of control are relatively low, since most data on the use of gas and the use of fossil fuels are available (MEZ 1992). Thus a certain number of 'fossil fuel use permits' can be identified to meet air quality standards. These may be distributed through grandfathering or may be sold.

Table 2.2. Share of the industrial sector in the emission of gasses.

Carbon dioxide	34.1%
Nitrogen dioxide	22%
Carbon monoxide	1%
Sulphur dioxide	65.3%

Source: IEA (1991).

As mentioned earlier, a distribution through grandfathering is politically more feasible. After the initial distribution, the participants in the market can sell, buy, rent or lease the permits. When a company is able to increase its efficiency in energy use, it needs less permits and the company can sell the permits.

A system in which the fossil fuel permits are traded without any restrictions may result in an increase of pollution in local areas. Therefore, some trading rules should be used to prevent adverse air quality impacts. With the help of zones in which different air quality standards are chosen, environmental authorities can control air pollution in urban areas. By setting different standards for different areas, the diffusion of pollution may be reduced, and a more compact city solution to the problem of urban sustainable development can be implemented. In non-attainment areas the trades might be limited in order to meet air quality standards for a local area. This might be done through an offset and a bubble policy, as was explained above.

The result of such a trading system is that - because of the restriction of energy use - the industrial sector is given incentives to use fossil fuels more efficiently and/or can seek cleaner ways of producing energy (e.g. through the use of renewable energy). A company will reduce the use of fossil fuels when the marginal costs of the reduction is lower than the price of a permit on the market.

2.5.3 The commercial / residential sector

Residential and commercial sectors use energy mainly for space heating, water heating, lighting, and more specifically, in the residential sector, for refrigeration and cooking. In 1991, the residential sector (in OECD countries) used about 18.8% of the total energy demand (IEA 1994a). More than half of the energy is used for space heating.

The number of consumers is large and the amount of energy use per consumer is relatively low. When the distribution is done through grandfathering, the actors in the market need to trade in order to arrive at an efficient allocation of the permits. The transaction and administration costs for this permit system will be too high in relation to the number of permits to be traded. It is thus not very likely that after grandfathering the permits, an optimal allocation will take place.

Another possibility is that the permits are sold through auctioning, so that the actors in the market do not have to trade because the initial allocation is already efficient. But also when the permits are distributed through an auction, the transaction costs will be high because of the large number of participants. The costs for consumers will be too high to make the system politically feasible.

The permits however, do not necessarily have to be bought by the consumers of the fossil fuels. They can also be bought by the producers and importers, who often have a regional or urban orientation and can raise the prices for the fossil fuels to account for the permits bought. The number of producers and importers of fossil fuels is rather low (in The Netherlands 40 to 50), and the transaction costs per permit will be lower. Secondly, the costs of control will be much lower. The environmental authorities only need to control the producers and importers who hold the permit (MEZ 1992). Since with this system the price for fossil fuels will rise, the consumers will receive clear incentives to economize on energy use. A difficult element in this system is to set the price that the consumers have to pay extra for the fossil fuels that they use.

2.5.4 The transport sector

The transport sector plays an important role in Western society. Transport allows personal mobility for both work and leisure activities. Transport also provides a vital lubricant for trade and has enabled the advantages of geographical specialization in production to be more fully exploited (Button and Rothengatter 1993). These advantages however, are also the cause of many environmental problems. Many people regard the transport sector as the main contributor to environmental problems in the urban area. The transportation sector is responsible for about 28% of the global carbon dioxide emissions, about 54% of nitrogen dioxide and about 90% of carbon monoxide emissions. In 1991 the transport sector was responsible for 30.8% of the energy consumption, up from a level of 23.8% in 1970 (OECD 1995).

As in residential and commercial sectors, the number of consumers in the transport sector is very large. The amount of fossil fuel per consumer is again relatively low. Grandfathering of the permits in the transport sector, has thus the same disadvantage as in the residential and commercial sector. Also when the permits are sold the transaction costs will be high, because of the large number of consumers (see also Vleugel 1995).

A possible solution would be to use the same system as suggested above for the residential and commercial sector. The permits might be sold to the vehicle producers and importers. Consequently the prices of fossil fuels would increase, because the producers and importers would have to pay for these permits. Trade of permits in the transport sector will then not take place among car drivers, since these users are not the holders of the permits. For a further discussion and overview of tradable permits in the transport sector we refer to Verhoef *et al.* (1997).

2.5.5 An integrated energy permit market

As argued above, it is difficult to establish a permit market in residential / commercial sectors and the transport sector on the basis of a permit for the fossil fuels used by the consumers. Nevertheless, it may be possible to create one overall fossil fuel permit market for the three sectors distinguished above. As mentioned, the costs of control and transaction are too high with many participants using relatively small amounts of fossil fuels. In the industrial sector we do not have these problems, so that permits in this market may be distributed through grandfathering, leading ultimately to a minimization of the costs of abatement. Since the transaction costs in the other two sectors will be high in relation to the amount traded, the result of trading will not be that the costs of abatement will be minimized.

A possible solution may then be for the permits of the two other sectors to be sold to the producers and importers at the level of initial distribution. The permits distributed through grandfathering in the industrial sector can then also be sold to the importers and producers of fossil fuels, when the industrial company has to buy fossil fuels from them. A big advantage of such a system is that the environmental authority only controls a limited number of permit holders, namely the producers and the importers of fossil fuels. In The Netherlands this would mean that approximately 35% of the fossil fuel permits would be distributed through grandfathering and the rest would be sold to importers and producers (MEZ 1992). In most industrialized countries, the energy used by the industrial sector is approximately the same. As said before, a trading market in the industrial sector will result in an efficient use of fossil fuels and/or the use of renewable resources. But also the importers and producers of fossil fuels will have these incentives. When these actors are able to reduce the fossil fuels in the residential/commercial sector and the transportation sector, they can sell their permits to the industrial sector. Information provision on energy saving techniques to the residential sector is important in this respect. Of course, the industrial sector might also sell permits to producers and importers, when it is cheaper for the industrial sector to reduce the use of fossil fuels. Clearly this is something else than giving the permits to the producers and importers when the industrial company buys the fossil fuels. In conclusion, there are interesting options for market-based urban energy policies, but for the time being these options are rife with many uncertainties and risks.

2.6 Conclusions

The experiences with urban sustainability policies so far are rather limited. Clearly, many cities have introduced urban environmental policies, but mostly they are of an ad hoc nature and not integrated in an overall balanced urban development perspective. In the present chapter we have addressed the question whether and how physical planning regulations and market initiatives may favour the attainment of urban sustainability. Although there is, in principle quite some potential, reality is somewhat harsh in that the quality of the urban environment is a multi-faceted and interwoven phenomenon, which cannot easily be treated in a comprehensive way.

And consequently, most policy initiatives are rather fragmented. Nevertheless, there is still some hope for more integrated urban initiatives, but these would require sufficient institutional support with a view to the potential offered by creative actions of stakeholders in the urban area.

An important aspect of sustainable development is that energy resources should not be overspent in order to offer future generations the chance to meet their welfare needs. A fossil fuel permit market seems a promising solution for both the problem of air pollution and the use of fossil fuels. It provides good incentives to favour renewable energy and to use the fossil fuel more efficiently. Such a system may be implemented at various levels ranging from the urban to national scale. The system may be made flexible with the help of different air quality standards for different areas, and with the help of offset and bubble policies, in order to prevent the spread of pollution activities to relatively 'clean' areas.

In conclusion, the system of marketable permits for achieving urban sustainability looks rather promising. In this context, there is certainly a need for more pilot studies in order to build up a body of knowledge on the potential role of market strategies for urban sustainability. In the empirical part of our study - concerning sustainability policies for European cities - we have unfortunately seldom found examples of successful applications of tradable permit systems. On the contrary, much emphasis appears to be placed on land-use and zoning principles, while it is noteworthy that the expected solutions of new energy/environmental technologies are rather optimistic. Many policy initiatives are related to technological innovation, which will be the main focus of the theoretical and empirical parts of this study, in particular in Part B.

3 Methodological Framework for Urban Energy Policy Analysis

3.1 Aims of the Study

Energy issues are an important policy concern. Not only is energy an input factor whose costs have to be minimized to maximize competitiveness in the industrial sector or welfare in the household sector, but energy use also plays a critical role in the solution of environmental problems. Consequently, in many EU countries there is strong interest in the effectiveness of energy policies.

Given that a concentration of activities implies a concentration of energy supply, urban areas seem to be a suitable geographical entity as a focus for energy policy. It is certainly true that the majority of Europeans live in urban areas, and that makes these areas suitable as focal points of energy-environmental policies. In the context of wise energy use, much attention is nowadays given to *renewable energies*. Renewable energy provision may offer a significant contribution to an improvement of local or regional environmental quality conditions. As a result, a new initiative focusing on the intricate relationship between renewable energy alternatives and local sustainability conditions is no doubt warranted.

Such a policy strategy may comprise various energy policy options, e.g. industrial cogeneration, district heating, combined heat and power (CHP) generation, combined urban waste treatment and energy production, peak load management, and institutional reforms in the structure of utilities.

In light of the above observations, a preponderant objective of the present book is to investigate the potentials of renewable energy policies in terms of benefits and of feasibility in urban areas. Such an ambitious goal requires the development of a methodology for identifying, assessing and evaluating critical success conditions for urban sustainability policies. The empirical application of such a methodology proceeds in our study through a cross-national comparative study of urban 'greenhouse' energy policies in various types of European cities, based on a systematically structured expert-based indicator list. The final outcome of this field experiment is a series of policy lessons based on qualitative and quantitative analyses, with a view to creating a better exchange of experiences between cities investigated, such that these lessons can be transferred to other cases. This above objective is put into operation on the basis of a series of steps:

- the *identification of potential critical factors for sustainable city policy* (using the so-called 'Pentagon' approach)
- the design of an *operational evaluation framework for alternative options for sustainable cities* by using the 'Pentagon' prism as a tool for comparative analysis of renewable energy options in cities
- a *field test on the above framework on the basis of a selected number of European (Dutch, Italian and Greek) urban sustainability policies*
- an analysis of the extent to which the relationship between *geographical structure* of cities and *levels of energy consumption* has implications for *urban sustainability*
- an *investigation into the driving forces or barriers faced during the planning and implementation stages of urban renewable energy policies* by using expert judgements on the viability of new energy options
- the *development of policy principles and guidelines on the question how and to what extent cities can make a significant contribution to coping with global warming problems* from the perspective of new renewable energy initiatives.

The empirical focus of the research is on renewable energy initiatives in various European cities for achieving a sustainable urban development. Clearly, the role of renewable energy differs for different cities in different geographical locations (e.g. solar and thermal energy in Greek cities, biogas and district heating in Dutch cities, or CHP in Italian cities). Several cities in Europe have been remarkably successful in some respects (e.g. Odense, Mannheim), but others are lagging behind (see for an overview Nijkamp and Perrels 1994). Therefore, a sample of 12 cities, equally distributed over 3 European countries (viz. Greece, Italy and The Netherlands), have been selected in order to extract strategic lessons on sustainable city policies based on a contrast analysis of these European cities.

Part C of the book essentially comprises a comparative evaluation of the potential benefits, weakness and strength factors of a number of renewable energy policies. It provides also a methodology for evaluating energy policies at the urban level, based on a set of critical success factors derived from the so-called 'Pentagon' model. This methodology is used in an extensive qualitative and quantitative analysis, in order to examine whether there are significant factors contributing to the rate of success of sustainable energy technologies in the above range of European cities. This analysis is based on empirical findings from consultations of key persons involved in initiatives in a series of case cities in the three countries concerned.

In Section 3.2 the logic behind our analysis is presented in more detail. Our main idea is that the adoption of energy policies in urban areas does not depend exclusively on technological barriers. Other - even more stringent - impediments do often exist which are of a different nature (organizational, financial, economic, ecological or institutional) and which often hamper often the acceptance mechanisms of energy policies.

A particular kind of promising but often undervalued energy policy which may be introduced in many cities is the adoption of renewable energy technologies. However, also the adoption processes of these technologies are hampered by barriers which go well beyond the technological level of maturity of renewable technologies. This is the subject matter of Section 3.3.

Section 3.4 then introduces the institutional changes faced in the energy sector, which inevitably influence the mechanisms associated with the adoption of these new technologies, e.g. by changing the level of competition among technologies and their comparative advantage.

Finally, Section 3.5 presents the relevant technologies envisaged and the cities involved in our empirical comparative analysis.

3.2 Research Strategy: the Pentagon Approach

The design, selection and implementation of various desirable renewable energy strategies for a sustainable city is partly an analytical and partly an empirical question, which may in principle be solved by using various assessment techniques and sophisticated (and nowadays computerized) evaluation methods. However, the choice in favour of a certain sustainability strategy does not necessarily mean that this particular option can easily be implemented, since there may be many obstacles (i.e. high transaction costs) preventing an efficient implementation of an environmentally-benign renewable energy system in a city. Several renewable energy policies may have a 'double dividend' character, i.e. they may favour environmental quality and may also reduce the costs of energy consumption. The advantages of the 'double edged sword' of such a policy may be offset by significant transformation costs (both financially and institutionally). In this regard, a trade-off between benefits and costs involved is necessary. While advantages of an urban renewable energy system are amongst others an improvement of local environmental quality, a reduction in (global) greenhouse gases and a cost saving in energy use, the costs of the installation of such a new energy system consist of direct investment costs in technical equipment and network adjustments and transaction costs due to inertia in institutions and among citizens. A basic aspect of investments in renewable energies is the lumpiness of the related capital stock, causing irreversibilities and thus high capital investment costs, which encourages a tendency towards risk avoidance among urban governments (Henderson 1981).

It should be added that the assessment of urban renewable energy strategies should not only take place on the basis of direct technical (engineering) criteria, but also on the basis of their overall contribution to urban welfare, income and employment. Thus, what is needed is a broad socio-economic evaluation (cf. also Nijkamp and Pepping 1998).

In this context, the use of the so-called Pentagon prism (see Figure 3.1) is meaningful, as this approach highlights the critical success factors (or failure factors) which act as necessary - though not sufficient - conditions for a successful introduction of a new technology. This approach has demonstrated its validity in various European infrastructure policy analyses (see for a detailed exposition (Nijkamp *et al.* 1994). Here, only a concise explanation will be given.

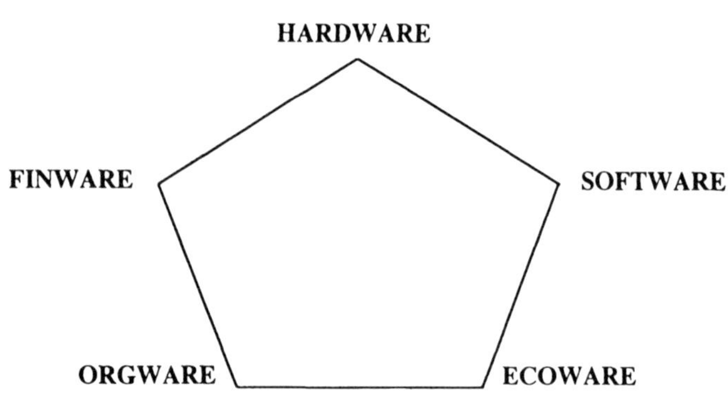

Figure 3.1. The Pentagon prism of critical success factors for the introduction of renewable energy technologies

Hardware factors refer to the level of technological sophistication of the renewable energy system used; for instance, a waste incineration plant with a high degree of heat recovery to be fed back into the public electricity network. Clearly, such investments require many resources and hence may face many barriers, which can only be overcome if one adopts a long-term viewpoint on urban sustainability as a *sine qua non* for long-range continuity.

Software factors refer to information provision and communication to citizens in order to induce environmentally-benign behaviour; for instance, an advertisement campaign to use energy-saving bulbs for household and office lighting. This requires pro-active and anticipatory policy strategies, since otherwise short-term policy and socio-political barriers will be too strong.

Next, *orgware* concerns the institutional and managerial efficiency in the urban energy-environmental sector; for instance, the introduction of market-oriented regulations for competition in the energy sector (e.g. third party access to the electricity network infrastructure). It also requires a sufficient integration of various urban policies, e.g. land use and transport.

The next success factor is *finware*, which is concerned with the cost saving and financing aspects of new energy initiatives; for instance, a financing scheme to stimulate the use of PV systems in urban energy provision. This factor is especially important for investments with a long pay-back period.

And finally, we have *ecoware*, which deals with the urban social and quality of life conditions for the implementation of new energy initiatives in a sustainable city context.

All these five critical conditions may influence the implementation process and potential success rate of renewable energy technologies. They may act as potential driving factors to eliminate various bottlenecks for the benefit for the environment of the city. For each of these five factors a series of issues can be identified which are more or less quantifiable. It is then possible to identify for each of the above mentioned five success factors a set of appropriate scores which may be given by experts as an approximation of the degree of success of a new renewable energy technology. In this way, alternative energy options can be compared from different perspectives and for different cities. These issues are related both to the *supply* and the *demand side* of technological applications (see Table 3.1).

It is clear that the above approach uses qualitative and semi-quantitative assessments of relevant factors. Therefore, it may sometimes be useful to resort to appropriate visualization methods and statistical methods for the comparative analysis of the technologies. Also modern meta-analysis is helpful in our comparative study (see Part C).

Table 3.1. Taxonomy of critical success factors for the adoption process of renewable energy technologies

	SUPPLY SIDE ASPECTS	DEMAND SIDE ASPECTS
Hardware	• R&D activities • Technological know-how - Technological know-how of other relevant technologies • Climatic conditions - Role of technology - Minimum efficient dimension - Compatibility with existing technologies	• Low relevance
Software	• Low relevance	• Public awareness • Adaptation of households • Adaptation of large users • Reliability for end-users • Cooperation in use • Participation of large users • Critical mass of users
Finware	• Initial governmental subsidies • Incomplete depreciation old technologies • Sunk costs • Profitability for supply actors	• Financial participation of users • Profitability for households • Profitability for large users
Orgware	• Organizational cooperation of supply actors • Division of responsibilities • Adjustment costs of management • Market conditions • Timing of introduction • Managerial learning processes • Requalification of existing staff	• Low relevance
Ecoware	• Saving of non-renewable sources • Reduction of pollution	• Visual disturbance • Harmony with public space functions • Harmony with building functions • Environmental awareness • Non-tangible social benefits

Caution should be taken with the transferability (generalizing) of results. Basically, a comparative evaluation of local cases of renewable energy applications does not immediately lead to the possible transfer of results from one (existing) case to other (not yet existing) cases. This is due to the fact that there is a large diversity in renewable energy policies. By definition, a renewable energy policy may be described as the combination of the application of a specific technology, the mixture of the organizational and financial structure and the promotional activities. This means that in fact, the setting of each application in each city is unique in this sense.

Nevertheless, these are also factors of a more general nature which have to be discussed as a general framework for a comparative study. These factors are in particular the energy production sector (Section 3.3) and the role of institutions (Section 3.4).

3.3 Adoption of Renewable Energy Technologies: the Perspective of Energy Producers

As mentioned in the previous sections, the broad aim of our analysis is the study of barriers and bottlenecks in the adoption of energy policies. In this respect, barriers and bottlenecks may arise not only in the field of policy-makers and policy decision processes, but may also stem from a rational calculation, based e.g. on a cost-benefit analysis, made by energy producers representing a substantive category of renewable technology adopters. In a static perspective, under the assumption of rational behaviour of economic agents, producers are inclined to adopt the new energy technologies when the adoption revenues exceed the costs, i.e. when a profitability margin exists for the producers.

In a study on energy policies, much attention has thus to be devoted to the decision-making processes regarding the adoption of renewable technologies by energy producers, who may affect the rhythm and the speed of adoption processes on the basis of profitability goals. In a dynamic perspective, the aim of a producer is not only to achieve profitability, but to adopt the new technology when the maximum profitability from the adoption is achieved. In this perspective, the problem of energy producers is to decide the right moment in time for the adoption.

Part B of the volume deals with the adoption of renewable energy technologies from the dynamic perspective of energy producers. The main problem in the adoption of new technologies for energy producers is that the adoption of renewable energy technologies implies a radical change from previous conventional energy technologies; it implies a break in the internal (micro-economic) learning processes of adopters (learning-by doing, by using, by searching, and indirectly, learning to learn) (Arthur 1988), as well as in the macro-institutional learning at the aggregate social level.

Technological change implies the generation of a new technological and organizational trajectory, based on new competence at both the micro level of adopters (the energy producers) and the customers (the energy users); the same changes are required at the macro level of the society at large. All this means that the introduction of renewable energy technologies implies a change in the *technological paradigm* à la Freeman, governing the economy and the society, a break with the already existing learning processes, and the start of new patterns of knowledge. For its cumulative and path-dependent nature, the development of a new paradigm is not guaranteed by the technical superiority of the incorporated technology. It is also necessary for the discounted value of the differential earnings expected from the new technology to be higher than the costs which have to be faced to bring the new technological trajectory into line (Camagni and Cappellin 1984). In this perspective, the discounted values of the different categories of costs and advantages over time allows to highlight the point at which the supply is profitable from an economic point of view.

Part B of this book presents a theoretical model of technology adoption in the light of the above observations: it is a dynamic model, where time is interpreted not in a strictly temporal dimension, but more at a conceptual level. In fact, in our model time is the rhythm with which innovation takes place, with which a new technological trajectory based on new knowledge starts, giving rise to new learning processes and cumulative trajectories. The choice of introducing a new technology may become an irreversible choice, due to cumulative investments which take place (David 1985). However, in the spirit of our approach, even a choice of non-adoption may become irreversible. Cumulative knowledge on the old technologies may change into exit barriers from the existing technological trajectory, in the form of sunk costs and complementary assets which reinforce themselves while time passes, finally turning profitable choices into non-profitable solutions.

The aim of the model is to present such irreversible mechanisms of non-adoption under different institutional regimes. Chapter 5 develops the model under monopolistic assumptions in the energy market, while Chapter 6 provides the same model under a more competitive structure. This difference reflects the institutional changes faced by the energy market at a European level, moving towards a more liberalized market regime.

The analytical results obtained from our theoretical exercise are tested in an empirical model in Chapter 7. On the basis of the database collected in our twelve European cities on bottlenecks and barriers of technology, a logit model is developed and tested with the aim to explain the reasons behind the adoption level and the willingness to adopt these technologies in the future.

After an exposition on the relevance of energy technology adoption, we will now present a concise overview of the institutional changes confronting the energy market, which explain also our interest in a more liberalized market structure.

3.4 Institutional Changes in the Energy Industry

The supply of a universal service like energy has historically justified the existence of a national carrier, for instance, the telecommunications sector, which operated under a monopolistic regime in both the supply and management of the system. A monopolistic structure is in fact able to guarantee the supply of a universal service, despite the real cost of supplying it, thanks to cross-subsidies among categories of users. One single national carrier operating in a monopolistic market can in fact guarantee:

- *continuity* in the provision of the service;
- *a universal service* provision;
- *equal tariffs treatment* for each category of users, despite the real costs of supplying the service to each category.

These principles are at the basis of a monopolistic structure in the energy market. They guarantee unbiased access to potential users, and tariff compensations, essentially of a geographical nature. This system implies that within a certain area of energy distribution, the national carrier applies the same prices to all users.

There are reasons, however, to think that in the long run a monopolistic structure cannot be justified even in the supply of a universal service. As discussed during these years, the impression is that these principles may be guaranteed with other mechanisms, than the monopolistic market structure (De Paoli 1995). The continuity in the provision of the service can be guaranteed by a governmental agency which has control over the adequate supply of the national production market. The principle of the universal service is related to the public nature of the service, which has to be provided to everybody for the sake of national welfare. However, once infrastructure has been developed at national level, as in the case of electricity grids, universality is by definition already guaranteed. Also the third principle may be obtained with other mechanisms than those of the monopolistic structure. For example, a balanced system regulated by a governmental agency can be put into place on the energy market, without being obliged to have only one national carrier.

For these reasons, the traditional monopolistic structure has been revisited in all European Countries, where different laws have been introduced with the aim to revise the institutional rules governing the market. Moreover, also at European level reflections on the best market structure have led to a European Union directive, which represents a clear step towards a more competitive structure in the energy sector. However, as usual, while these institutional changes are introduced to overcome certain market distortions, they inevitably introduce other distorting mechanisms, at least in the transition phase. Inevitably, these changes influence the adoption process of renewable technologies, since they influence adoption profitability and its development over time, as we will see in Part B of the present book.

The institutional changes affecting the energy sector are varied; they are presented in this section in the case of the three countries, The Netherlands, Greece and Italy, for which the empirical analysis is carried out.

3.4.1 Institutional changes in the energy market in The Netherlands

In The Netherlands energy companies and the government both have important roles in achieving a reliable and clean energy supply. It is acknowledged that a good operation of energy markets is necessary, and the energy policy is based on interventions of the government only when this is necessary in order to achieve the envisaged energy targets. Governmental interventions take place in three policy areas:

- constitution of conditions for energy markets (competition, environment, safety, urban planning)
- specific care and interest related to energy (savings, renewable sources, risk spreading across fuels and suppliers)
- control of national resources from the soil, mainly the natural gas reserves.

The markets for oil and coals are international, with mainly large suppliers. Governmental intervention is restricted here to general issues concerned with competition, environment, urban planning and safety. For the electricity and gas market the situation is different. The actual electricity market is based on the Electricity Law 1989. This law imposes on the energy sector, amongst others, regulations regarding market structure. The gas market, in contrast to the electricity market, has no laws regulating its organization. Insofar as laws exist for this sector, they have only minor influence on the organization of gas supply and are mainly directed to the supply side (mine law). There are private agreements between the government and the gas sector, in which Gasunie (The Dutch gas supplier) assumes a crucial place. Thus, both with electricity and gas there are market-like elements, but nowadays the government realizes that the developments in these markets demand a new institutional framework.

When we look at the electricity market, as a result of the 1989 Electricity Law, market conformity has already been improved. There is a clear, historically grown, monopoly. The striving for economic scale advantages led to a single supplier for each region. In that period the emphasis was on the insurance of supply with a rapidly growing demand. After the oil crisis in 1974, this situation changed. Initially the higher energy prices led to a larger emphasis on the risk diversification of fuels and a restrictive use of gas, while maintaining the monopolistic structure. In the eighties, the awareness grew that with a view to higher energy prices the perspective for CHP and decentralized energy technology development improved, and that the monopolistic structure was not suited for its development. This motive, together with the need for cost reduction and higher savings in energy use, led to more political pressure to introduce free competition.

The Electricity Law 1989 contains a large number of stimuli for competition. Most important is the freedom of decentralized power generation. Electricity surpluses may be sold to distribution companies for a relatively high price, and the latter were obliged to buy. Furthermore, the generation of more than 25 MW by joint ventures of energy companies with private bodies was left to the free market. The new law also contained many market-based measures giving more freedom to large-scale users and distribution companies. Freedom of export and import (the latter not for distribution companies), transport rights, and the possibility to buy from other producers. However, until now only a modest use has been made of these facilities due to small price differences, high transaction costs and the fact that Dutch energy prices are internationally attractive, also because of low gas prices.

The 1989 Electricity Law has resulted in a rapid introduction of decentralized CHP systems, nowadays supplying an important share in total energy. In addition, the law has led to a rationalization of energy company strategies and more uniformity and transparency of tariffs.

In the Third Energy Memorandum of the Dutch government (1995) (Tweede Kamer der Staten-Generaal, 1995a), the planning for institutional changes in the energy market was set for the years after 1996. Compared with the actual situation, there have been some major changes which can be grouped into main issues, which apply to both the electricity and gas markets to a lesser or larger extent.

Firstly, the energy markets that depend on pipeline or grid networks would, like other energy markets, no longer be dominated by supply, but by demand. The initiative is taken by the users with the consequence that their number of choices increase. This new concept means that 'network functions' must become less dependent on other functions, like production and distribution of energy. Therefore, the entry to the network must be non-discriminatory for potential newcomers on the market. This development applies more aptly to electricity and gas networks than heat networks, as at the moment a more market-based development of concentrated heat generation is not yet possible from a practical point of view. This new concept also means that governmental care for energy supply in networks can concentrate on 'captive users', i.e. energy consumers connected to networks, who have no choice regarding the type of energy source.

The Dutch government maintains its regulatory influence on both the energy networks and the protection of captive users. The EU treaty gives (via Article 90.2) governments the possibility to instruct firms to fulfil tasks for general interest. When necessary, these firms are then exempted from stipulations in the competition regulation law. Also the Dutch Law on Economic Competition supports these possibilities, as well as the newly proposed European gas and electricity directives.

The 'choice-users' are essentially those users consuming annually more than 10 million kWh or 10 million m^2 of gas. These users have a freedom of choice regarding the source of energy. This measure made 30% of the electricity market and 40% of the gas market free on the demand side. Furthermore, there is a smaller group of large-scale users who will have free choice at the latest after the year 2001. This covers another 30% of the electricity market and a further 20% of the gas market. However, small private users remain 'captive users', who are dependent on existing energy suppliers. The Dutch government keeps protecting these users against possible abuse of power by the suppliers. This protection concerns supply security and supply conditions (price and quality). The suppliers have an obligation to supply. Therefore, they have to make provisions to ensure the supply, and commit themselves for longer time periods to other producers. The consequent necessary commitment of users means exclusive supply areas for these energy suppliers.

Dutch energy companies are extending their activities increasingly beyond the energy market. These are markets with close connections with energy (for example, energy services), sometimes also public utilities (waste management and the environment), but also with commercial activities like telecommunication. This extension of activities has a positive influence on the quality and efficiency of the activities in the energy field. The disadvantage is that it generates financial risks that may be shifted to captive users, which in turn may disturb the competition position relative to other market parties. The government therefore regulates such activities in the following way:

- new activities have to conform to the market, i.e. they may not affect the free competition on the markets on which they operate;
- new activities may not lead to disadvantages for non-choice users.

The first of these targets is covered by a new arrangement in the Law on Energy Distribution (Article 23i) which creates a system of obligatory separation of different activities. It facilitates the treatment of competitive forces in a legal procedure. The same article supports the second aim by giving user-cooperations the task to monitor possible effects of new activities on the energy supply and on the energy tariffs for captive users. It is expected that the two above mentioned government tasks relate to a decreasing group, as the competition between energy suppliers will increase (and with that the share of choice users).

At the moment, most Dutch energy companies in the public sector are still owned by the provinces and local public energy companies, but there is a tendency towards increasing liberalization and privatization of the energy supply markets. However, as large regional energy companies have a special function regarding the economic development of regions, shareholders often want to keep control over their strategies. Furthermore, privatization is taking place more rapidly in markets with sufficient levels of competition like the electricity market. The Dutch government stimulates the subsequent introduction of liberalization and privatization as, according to the processes in other countries, a simultaneous introduction of both processes may create private - instead of public - energy monopolies.

The above described developments also have implications for energy companies in a fiscal respect. Corporations in which non-energy related activities are lodged will become subject to corporate taxes. The above mentioned article in the Law on Energy Distribution therefore also arranges an adjustment in the Law on Partnership Taxes 1969. Furthermore, by means of a law change, energy companies carrying out those activities will become partly subject to corporate taxes as these activities are concerned because of increasing levels of competition.

We will now discuss in more detail the institutional framework of the innovative energy technologies CHP/DH and renewable energies. The development of CHP in The Netherlands has been successful. CHP remains attractive because of the high quality of Dutch energy infrastructure for both gas and electricity, and the good supply possibilities in the energy intensive industries, horticulture and cities. In the year 2000, the total CHP capacity will be around 8000 MW, which is 40% of the total capacity. A bottleneck is created by the fact that heat demand and electricity demand do not always have parallel patterns. It remains possible to provide CHP heating to the most profitable parts of the heat market.

The largest autonomous grower is the industrial CHP sector. Industries tend more and more to implement this technology to cover their demand for process heat, mostly in cooperation with production and distribution companies. The generated electricity is sold to electricity companies or other market parties. Industrial CHP appears to be profitable and there is no need for the Dutch government to intervene on this market.

The situation is more difficult for CHP centralized systems with heat networks (urban heat networks and networks for horticulture). These initiatives bear high financial risks that are related to the distribution of heat (see also Chapter 8). This is not in line with a more market-directed orientation of this sector. Because heat networks are part of the infrastructure for more sustainable energy supply, fiscal facilities are applicable to this sector.

Finally, we will discuss small-scale CHP. The developments resulting from the 1989 Electricity Law have shown that it was impossible to maintain the favouring of decentralized power generation. According to this law, decentralized electric power has to be offered at a non-market value since this form of energy has a risk component different from centralized power generation. However, in recent years, the share of decentralized power generation has grown to such an extent that nowadays there is an over-capacity in this sector. In addition, CHP has taken such a prominent role in the market that new centralized power generation in most cases also includes CHP. The difference between centralized and decentralized power generation is therefore diminishing and no longer calls for selective consideration by the government.

In order to speed up the implementation of innovative energy technologies, the Dutch government uses several stimulating measures. The following measures are already applied:

- a special arrangement for renewable energies in the regulating energy tax system which came into force in 1996;
- *green current*: an initiative of the energy distribution companies that can stimulate the application of renewable energy and also enlarges the awareness and involvement of the public during further implementation;
- *green investment*: an existing facility that reduces interest charges of investments in renewable energies (Tweede Kamer der Staten-Generaal 1995b);
- an agreement between Dutch wind energy producers and the overall organization of regional energy distribution companies on the compensation for supply to the network of electricity generated by windmills;
- initiatives of the energy distribution sector to stimulate further penetration of renewable energies in the framework of the national Environmental Action Plan (MAP) (Stuurgroep Integraal Milieubeleidsplan Energiedistributiesector 1991).

To achieve the national target that by the year 2020, 10% of the energy demand will be covered by renewable energies, another supporting programme has been initiated called Action Plan Renewable Energy. This programme consists of five parts. First, there is an extension of the fiscal measures with a positive influence on the profitability of renewable energies. These consist mainly of tax credits on investments and reduction of the value added tax tariff for 'green current'. Secondly, there is an enlargement of the budget for R&D activities with 25 million guilders per year for the whole range of technologies. Thirdly, the budget for demonstration projects and market introduction of renewable energy technologies has been enlarged by 25 million guilders. This is subdivided as follows among the different technologies: biomass 5 million guilders, photovoltaic solar energy 15 million guilders (partly for market introduction and partly a more substantial contribution, mainly to reduce the cost price of this energy form), and wind energy 5 million guilders. In the fourth place, the Action Plan Renewable Energy guarantees an adequate compensation for supply to the network of electricity generated by renewable sources by private firms or individuals. Finally, there will be a continuous monitoring of the use of renewable energies. If all this is not achieved, a minimum share of renewable energy in the supply to captive users can be enforced.

3.4.2 Institutional changes in the energy market in Greece

The traditional institutional framework that governs the production and distribution of energy in Greece is characterized by the dominant role of the Public Power

Cooperation (PPC) company. It is a public entity that exclusively produces and distributes electric energy.

In this context, the production and consumption of electricity took place in a non-competitive framework where all decisions were taken by public authorities. This institutional model incorporated several positive and negative aspects. In the first place, PPC distributed the energy all over the Greek territory, thus offering a considerable socio-economic contribution because of the characteristics of the Greek landscape that imposes serious isolation problems. Moreover, the exclusive role of PPC might lead to market disturbances and hence to a relatively inefficient energy use. On the other hand, for other sources of energy (fossil fuels), the production and distribution took place in a relatively competitive context. Public authorities intervened here only to correct market failures arising from the oligopolic framework.

As far as the renewable energy resources are concerned, their use was promoted indirectly and accidentally. Specifically, the renewable resources in Greece pertained to the energy forms that were mainly produced by PPC.

In the traditional institutional framework there was no specific legal context that dealt with the utilization of renewable resources. However, there were some laws, mainly concerned with economic and social issues, that impacted on their use, namely:

- *the Law 1962/1982*; this law concerned the regional development in Greece. In fact, it indicated a series of financial taxes, administrative and other incentives for investments that aimed to initiate regional development. In this context, it offered financial support that adds up to 15% of the investment when the aim is to produce energy from renewable resources. Specifically, the law envisaged the following energy production methods: substitution of fossil fuel and electricity by renewable resources or recycling of waste heat. We note that the law did not cover the new production of energy, but only the substitution of already existing production systems.
- *the act for installation of solar energy in households*; in the spirit of the above law, a specific legal act was passed for the installation of solar systems for water heating in households. Especially financial support was offered through this act which covered the entire investment of installing a solar system. This particular support took the form of long run loans with a rather low interest rate.
- *the 'Valoren' programme*; 'Valoren' was an EU programme aimed at regional development via the utilization of indigenous energy resources. 'Valoren' offered financial support of up to 70% of the relevant study. The Ministry of National

Economy decided that the 'Valoren' financial support could not be additional to the support of the 1962/82 Law. Therefore, for each 'Valoren' investment, the Ministry of National Economy withdrew the sum that the investment received via the 1962/82 Law. This led to considerably moderate results from the EU additional effort.

Pressures for a new institutional framework for the use of renewable energy resources are occurring, due to current pollution problems. A new institutional framework is now formulated via a number of national laws, some important pilot programmes and projects, and the promotion of a more competitive context for the production of energy.

Some other legal frameworks also play a role in the Greek context, in particular the 1512/1985 Law and Law 1559/1985. The first one is the oldest one pertaining to the new Greek institutional framework; it concerns the energy saving methods in heating and cooling of buildings. It is worth mentioning that the law was largely inactive and only in the last period has it been more strictly enforced. This law considers the following issues:

- standards for entire Greece as well as some specific standards for particular areas. These standards concern building materials, the installation of energy or passive solar systems, buildings insulation systems etc. The law indicates that those buildings based on these standards will receive financial and other support.
- standards for building location. The criteria of the location are oriented towards energy saving buildings according to the characteristics of each area.
- development of a pricing method for building heating expenditures that takes into account the thermal losses of buildings.
- application of financial incentives for the improvement of energy efficiency of central heating systems in existing buildings.

This general framework is further elaborated for various classes of buildings. Next we will concisely discuss the 1559/1985 Law, which essentially initiated an indirect endeavour towards energy production from renewable resources. Specifically, it indicates that the production of energy is permitted for autonomous producers if and only if the production utilizes renewable resources and the energy is being partially consumed by the producer; an autonomous producer may be interpreted as any private, public, or semi-public producer that does not belong to the Public Power Cooperation (PPC). PPC still maintains the exclusive privilege to produce and distribute electric energy. In this context, the autonomous producers can sell the surplus between their production and consumption, provided that their production is less than three times their consumption.

Evidently, this condition indicates that the law does not permit investments that are exclusively aimed at producing renewable energy for commercial purposes. Instead, it promotes the production of renewable energy from big consumers for their own use and as a side line they may sell part of the energy in order to support their investment. In this framework, the PPC should be the exclusive buyer of the energy sold by the autonomous producers. The price is determined by administrative decisions.

Strong advantages are given to local communities and municipalities, who may exclusively produce energy for commercial use, escaping from the constraints of self-production that are imposed on private producers.

The enforcement of this law has revealed a serious drawback. Some potential producers, mainly local authorities, have requested - and received - the formal permission for installing a production system and by so doing they blocked the relevant resources while they postponed or even suspended the investment. As a result, other potential users of that specific resource were excluded. It has been argued, that in order to cope with this problem a deadline should be imposed on each formal permission; when the deadline lapses, the permission will become invalid.

Finally, we will pay attention to Law 2244/1994. This law is the most recent legal action that concerns the utilization of renewable energy resources and energy saving initiatives. The new element of the law is the opening up of renewable energy production to private and public entities. The stipulation that a producer of renewable energy should be the main consumer and can only partially sell energy, is withdrawn by Law 2244/1994. Thus, the law permits the independent production of energy from renewable resources. However, the product can be sold exclusively to the PPC which is in charge of the distribution system of the electricity sector. The relevant prices are determined by administrative decisions and could differ in space and time. In fact, Law 2244/1994 initiates a considerable endeavour towards the competitive function of renewable energy production.

PPC remains the main body that is responsible for producing and distributing energy in Greece. Almost 99% of the energy production and distribution in Greece is carried out by PPC. In the early 70's, PPC devoted considerable effort to renewable energy resources. Initially, PPC developed a widespread measurement network and a number of advanced mathematical models for the evaluation of the potential uses of renewable resources. Based on these findings, PPC pinpointed a number of sites where the generation of electricity from renewable resources was economically sound on a commercial basis. Moreover, PPC was seeking to find adequate solutions to specialized problems associated with the energy demand in isolated areas, especially islands. In these cases, renewable resources seemed to be an attractive solution.

The above does not imply that PPC started a wide application programme for utilizing renewable energy resources. Rather, it initiated a pioneering and experimental programme. In the context of this programme, PPC cooperated closely with academic research institutes for solving several problems associated with the commercial use of renewable resources.

In general, nowadays much effort seems to be made from an institutional point of view to facilitate and support the adoption of renewable technologies in Greece, although the openness to new institutional bodies and energy producers is extremely limited. A different situation exists in Italy, the third country in our study, where a high degree of openness to the energy market is manifested, although fewer institutional rules supporting renewable energy systems have been put in place.

3.4.3 Institutional changes in the energy market in Italy

The institutional changes developed so far in Italy are linked only to a very limited part of the complex 'filière' of the energy industry. The phase where the monopolistic structure is under revision is the energy production phase, while the other backward and forward phases of the production chain are either already under a competitive or under an oligopolistic structure. This latter characterizes the case of the energy equipment production market, which is in the hands of a few large European firms, holding the necessary technological knowledge to produce in this market. On the contrary, the energy transmission and distribution phases are still under a monopolistic regime.

In the whole energy industry, the famous Law 9/10, introduced in 1991 in Italy, has broken the monopolistic structure of the energy production market allowing private energy suppliers to enter the market and to supply energy, with the obligation to sell the energy produced to the national carrier (Enel). The way in which this law has been conceived is peculiar: it is not the result of a general reflection on the appropriate market structure in the energy production sector, but the result of the need to cover a deficit in the national energy production of the national supplier. The national energy plans in 1988 revealed in fact an inadequacy of national energy supply which obliged the national carrier to open up the market to private suppliers. Seen from this perspective, the national monopolistic structure was no longer sustainable, and the openness of the market for private suppliers was the simple consequence.

Market entry takes place with an acquisition of a licence provided by the national carrier to the private supplier. Until a few years ago, the national carrier chose the entrants exclusively on the basis of the quantity of additional energy required to cover the deficit. The consequence of such a strategy resulted in a very fast market saturation, and in the closure of the market by the national carrier. Recently, the procedure to obtain the licence has been changed by the government and a technical, financial and operational feasibility study of the power plants' implementation projects is now required, where potential entrants specify the quantity of energy supplied, the technical characteristics of the plants, the location of new plants and the date of production start. Feasibility studies allow the national supplier to estimate the national energy production capacity.

In this law, the monopolistic position of the national carrier in the field of energy transmission and distribution is guaranteed. Private suppliers are in fact obliged to sell their energy produced to the national carrier, at prices decided by the inter-ministerial price committee (CIP) in the national Law 6/92. Prices vary according to the technology used to produce the energy.

The monopolistic market structure in the field of transmission and distribution will change in the next two years. By that time, European countries will be obliged to adjust the situation to the European Directive, approved in January 1997, which liberalizes the possibility to sell the energy directly to large users. The private supplier however, is obliged to pay a price of transmission to the national carrier. Each Member State is free to choose among possible alternative systems, and two of them have already been envisaged and applied in two European Countries:

- a system where a single buyer exists, and plays the role of a mediator between private suppliers and energy carriers. This model, already applied in France, works on the basis of public auctions in which private suppliers decide on a medium-long term contract with the carrier;
- a system of free access to the distribution network, as in the case of Great Britain. In this model, the market is controlled by a governmental agent (in the case of Great Britain, the National Grid), with a daily auction system.

These systems are very different: the first model is more concerned with a monopolistic market, where the entry is regulated by long-term contracts while the second model works more on competitive rules typical of an auction system, where the price element plays a fundamental role as the entry criterion.

In fact, with the Law 9/10 of 1991, the monopolistic position of the national carrier in the field of energy production is broken, at least in theory. The way in which the law allows for entry in the market does not lead the industry towards free competition. The entry is strictly controlled by the national producer whereas in a real competitive structure, the role of a market regulator should be played by a national authority, not involved in the market.

Another distorting mechanism in the current changes is linked to the more recent law governing prices to sell energy to the national carrier (Law 6/92 of CIP). The relative price advantages have paradoxically been fixed for technologies which are already in a commercial phase. Relative prices have in fact been fixed in favour of energy efficiency improvement technologies, like district heating based on conventional energy sources, thus penalizing less mature technologies such as wind and PV systems.

Finally, even the European Directive through the way in which it is conceived generates some distorting mechanisms: it provokes unequal tariff treatments among categories of users, and in particular between large users (large firms) and small users (residents and households), who are still obliged to buy energy only from the national carrier at the monopolistic price. If this is justified by the argument that the national supplier provides a universal service, some reflections are in order on the real need to protect the universality of the service, once the infrastructure provision is guaranteed on the whole national territory.

A question is legitimate at this point: which new market structure will really emerge? This question can be answered by summarizing the changes as follows:

- before 1991, a typical monopolistic market structure existed;
- after Law 9/10 of 1991, a competitive market structure characterizes the market of energy production. Competition takes place not on the basis of price mechanisms (energy prices are defined by Law 6/92 of CIP, and are thus equal to each supplier), but on production costs, through technical efficiency of power plants, of the technology used in general, and of organizations and managerial costs. In this framework, the position of local carriers, which historically own power plants and distribution grids because of a legal agreement between them and the national carrier in legally defined parts of urban areas, remains unchanged. Local energy public companies still retain their monopolistic power;
- after the European Directive is put in place, competition will be based also on price mechanisms, at least for the large users market.

3.4.4 Concluding remarks

The energy market is a complex one, both technically and organizationally. This market exhibits a high degree of dynamics in the past years, and will likely be in a state of flux in the years to come. We will in Part B present a theoretical model for the adoption of renewable technologies, under different institutional regimes. The model aims to analyse the way in which adoption costs and revenues change over time, and thus addresses the optimal time of adoption, where the profitability of renewable technologies is maximum. As we will see, adoption costs and revenues have different time trends, according to the institutional rules governing the market. Consequently, each market structure suggests a different adoption behaviour of firms. In this respect, through this exercise, we are able to answer the question whether or not the institutional changes taking place in the market facilitate the adoption of renewable energy technologies.

In particular, a dynamic adoption model is built under 'extreme' assumptions of institutional changes, so as to distinguish the different situations in a clear cut way. Three different cases will be analysed:

- a monopolistic structure, where the single producer may decide if and when it is convenient to enter the market of renewable energy technologies;
- an imposed cooperative structure, where more than one firm operates on the national market, but competition is based on costs rather than on price mechanisms;
- a competitive structure, where energy can be sold by private suppliers to final users at the desired price.

Since our study has a particular focus on adoption of renewable energy at the *urban scale*, the model, once open to competition, is in reality developed to define the choice of adoption made by local energy producers, which possess a very high share of the urban market. The model tries to determine the optimal time of adoption, when a local monopolistic supplier faces national competition, or national cooperation. In this context, Chapter 5 introduces a model under a monopolistic regime. Chapter 6 then provides a model of a local monopoly versus national competition, where the latter is both an imposed cooperative market structure and a real competitive market structure.

Before entering Part B and its theoretical analysis, it is important to discuss the technologies and cities chosen for the analysis of the barriers and bottlenecks in the adoption of renewable technologies, contained in both Parts B and C. This will be done in Section 3.5.

3.5 Typology of Energy Production Technologies and Case Cities

Clearly, most cities in Europe exercise in one way or another some sort of energy policy. In our research we have included interesting cases which may generate important lessons for other cities. The scale on which our European fieldwork is set up reduces the generalization problem described in the previous section. Our case study covers a series of twelve different cities in three different European countries, so that a bias due to specific organizations and financial settings is initiated. This makes it possible to use a typology of renewable energy policy options based on their key feature, namely the technology adopted.

To draw a general typology of innovative/renewable energy production technologies, we will adopt the following operative definition of renewable resources: sun, wind, hydraulic energy, geothermal resources, tidal energy, and the transformation of organic and inorganic refuses or refuses from vegetable products. Technologies for a more efficient use of energy are considered as combined heat and power generation: the use of heat recovered from waste smokes and from thermal plants, and the use of other energies recovered by processes, plants and products. Finally, there are innovative technologies for direct savings of energy: energy saving on climatization and lighting with policy interventions on buildings and plants.

Our empirical application is based on the above definition of innovative technologies, with emphasis on the use of renewable sources. However, some initiatives are not worth studying for our scope, because they do not (yet) exist in practical terms or because they are parcelled out with a very little dimension and therefore difficult to investigate. Others, on the contrary, have no relevance in terms of city systems of energy production or do not represent a real innovation in the field of energy production.

The technologies chosen finally to be the object of our work, make use of *biomass energy, solar energy, wind energy and energy saving*. These are the following (in brackets the often used abbreviations):

- Photovoltaic systems connected to an electricity network (PV)
- Solar water heating systems (SWH)

- Non-individual wind energy (aeolian) systems (CWE)
- Energy from refuse incineration (RI)
- Energy from the recovering of biogas from refuse (BIO)
- Combined heat-power generation with district heating (CHP/DH)
- District heating fed by renewable sources (DH/R)
- Insulation technologies for savings in energy use (SAV).

The last mentioned insulation technologies for savings in energy use will receive less attention, as they are not really concerned with a more efficient production of usable energy, but are more oriented towards the reduction of the final consumption of energy.

It has to be noted that PV and aeolian systems represent interesting cases of non-adoption (or of a very difficult adoption) in many cities. The study of these 'non-cases' gives us the occasion to understand whether and why the not so favourable local climatic conditions - emerging intuitively as an evident bottleneck for their introduction - are really the only factors obstructing their adoption.

Our comparative policy study concerns Dutch, Italian and Greek urban sustainability initiatives. These three countries were selected in the framework of an exploratory study commissioned by the European Commission. The choice of the particular case cities was determined by the aim to compare cities that were different with regard to their size, their political-organizational arrangements and the urban environmental initiatives carried out (e.g. different levels of renewable energy adoption). The participating cities, which certainly form a contrast with the above mentioned viewpoints, are presented in Table 3.2.

The Dutch cities chosen are *Amsterdam*, the capital city of The Netherlands which is now experimenting with a large-scale application of photovoltaic roof panels in one of its newly built areas; *Rotterdam*, a big mainport, being the first city in The Netherlands with a large-scale district heating system developed just after the Second World War; *Almere*, a medium-sized new town with a recent application of district heating, which has the highest population growth rate in The Netherlands, so that this dynamic city may be an interesting test case for new urban spatial development and integrated urban energy planning; and finally *Gouda*, the city playing the pioneering role in The Netherlands in the area of solar water heating applications (being the 'Solar Water Heating City No.1 in 1995).

The Italian case cities are *Milan*, one of the most attractive large business centres in the Mediterranean, with its recent methanization (which is apparently only to the detriment of a district heating system), its advanced differentiated waste collection system and its new Urban Traffic Plan; the city of *Turin*, the first city that developed an Energy City Plan and a large district heating system already deeply rooted in its territory; *Brescia*, an interesting and successful example of large district heating plants and integrated systems to supply energy; and finally *Pavia*, a small provincial city that is in the centre of heavy and numerous transport movements that are far beyond its natural capacity, and for this reason needs a general urban mobility plan focused on a reduction in energy use and environmental decay.

Table 3.2. Case cities in the three countries investigated

Country	*Large cities*	*Medium-size cities*
The Netherlands	Amsterdam	Almere
	Rotterdam	Gouda
Italy	Milan	Pavia
	Turin	Brescia
Greece	Thessaloniki	Mytilini
	Chania	Ptolemaida

The Greek study cases are *Chania*, a relatively large city on Crete where recently some initiatives have been started up by the Organization for the Development of Western Crete (wind energy and photovoltaic application, both in an early project phase); *Thessaloniki*, the second largest city of Greece, where some small-scale applications of solar space and water heating exist; *Mytilini*, the medium-sized capital city of the island of Lesvos with a windmill park; and finally *Ptolemaida* (and its neighbour town Kozani), that plays a pioneering role in Greece in the area of district heating.

In these cities, policy-makers, engineers and representatives from energy companies were interviewed by means of structured interview lists, which aimed to assess the relevance and successes/failures of the various city initiatives by using the Pentagon prism as an analytical framework. In addition, they were also asked to assign quantitative achievement scores for both the perceived rate of success of the respective technologies, and for the experienced (in the cases of existing applications) or expected (in the cases of not yet existing applications) extent to which the five Pentagon factors act as bottleneck or facilitating factors to the introduction or the viability of these technologies. This information has been systematically collected by means of a common checklist in all three countries (see Table 3.1), and a database of sixty discrete observations obtained (as a result of three countries, four cities in each country and four technologies in each city). Statistical analyses and discrete choice models (logit and probit models) are run on the basis of this database. However, the limited number of observations for each technology does not allow us to run a statistical analysis for each technology separately. For this reason, a more recently developed meta-analytic method, viz. rough set analysis, is applied in order to analyse each technology separately. The interviews run are also useful to provide a qualitative analysis, which provides an exploratory and explanatory insight into the weakness and strength profiles of the various technologies.

Having presented now the key issues in the adoption of renewable energy technologies as part of urban sustainability initiatives, we will in Part B address more explicitly the economics of the adoption of such technologies leading into our comparative analysis dealt with in Part C.

PART B

URBAN SUSTAINABILITY AND ENERGY TECHNOLOGY ADOPTION

4 Key Issues in the Adoption of Renewable Energy Technologies

4.1 The Need for Policy Initiatives

A significant increase in global energy consumption is likely to take place in the forthcoming decade. This increase cannot be met in the long run by fossil fuel based power supply, without very serious consequences for the global environment (Lundsager 1996). Appropriate environmental policies to keep environmental costs under control will be more and more necessary and they will have to be put in place through institutional rules monitoring and governing the behaviour of firms and individual energy users.

Within the wide range of proposed regulatory measures aiming to cope with the environmental damage caused by energy use, technological change towards more environmentally-benign technologies has been advocated as an efficient policy strategy. The exploitation of technologies based on renewable energies seems to offer a great opportunity in the near future (OECD 1995). This is an interesting option especially for urban settlements. It is widely recognized that cities are the main energy users and, consequently, the main producers of air, water and land pollution. In fact, most production and consumption activities of modern economies take place in cities, and hence they are the largest consumers of energy and materials. At the same time, cities are the most appropriate location for human settlements. A very large percentage of the world population lives in them, although sometimes quality-of-life and environmental concerns undermine socio-economic advantages associated with urbanization. Both the high urban population densities and the high energy use in cities have put the urban sustainability debate at the heart of the problem of control of energy use in cities.

Although an increasing use of energy in urban settlements is foreseeable in each industrialized country, a process witnessed by national policy initiatives aiming to develop laws and rules in favour of renewable energy technologies, the implementation and adoption of environmental-benign technologies is still modest in most industrialized countries.

A glance at some recent data on the penetration level of different technologies in different countries reveals that the level of adoption of renewable energy options is astonishingly low and varies greatly over many countries.

Table 4.1. Renewable Energy Produced in European countries

Countries	1990	1994
Austria	22.1	24.1
Sweden	24.7	24
Finland	18.9	19.3
Portugal	17.6	17.5
France	6.4	7.2
Greece	7.1	7.2
Danemark	6.3	7
Italy	5.3	6.4
Spain	6.7	6.2
Irland	1.6	2.2
Germany	1.7	1.9
The Netherlands	1.3	1.4
Luxemburg	1.3	1.3
Belgium	1	0.8
United Kingdom	0.5	0.6
European Union	**5**	**5.4**

Source: Green Paper, "*Energy for the Future: Green Paper of Energy*", EU, Brussels

Table 4.1 shows the share of energy produced with renewable technologies in 1994 in most European countries[1]. The share varies significantly around the European average (around 5% of the energy produced), ranging from 24% in Austria and Switzerland to about 0.6% in Great Britain. Even more astonishing is

[1] The data contain also energy produced with hydro-power plants.

the growth rate of renewable technologies in the period between 1990-1994. At European level, the energy produced with renewable technologies appeared to remain nearly constant. A similar trend applies to almost each individual country. Apparently renewable energy prices are not yet competitive.

When dealing with the penetration rate of renewable technologies using energy sources, leaving aside hydro-power energy plants and energy efficiency improvement plants, the situation is even more worrying.

Table 4.2. Energy Produced by Renewable Technologies in 1990 in Italy

Sources	Installed capacity	Thermal energy produced (IJ)	Electricity produced (GWh)	Total energy produced (Mtep)	Share
Solar	290000 sq meter	100		0.007	0.06
Photovoltaic	4,0 MWp		0.7	0.001	0.01
Geothermal	500 MWe	8,400	3,100	0.9	7.63
Biomass:					
* Wood		117,000	23.5	2.81	23.81
* Waste		2,350	85	0.075	0.64
* Biogas		121,700	153.5	2.9	24.58
Wind energy	2,9 MWe		2.45	0.0006	0.01
Hydroelectric	18,966		35,079	7.91	67.03
Total		133,000	38,339	11.8	100
Share on total energy produced				7%	

Source: Cavaggioni (1992)

Table 4.2 shows that in Italy among existing renewable technologies, hydropower plants produce more than 67% of total energy produced with renewable technologies. Solar energy and photovoltaic systems produce a very limited share, around 0.06% and 0.01%, respectively, for solar and photovoltaic plants. A similar situation applies to more advanced countries like the US, which shows a high penetration of hydro-power energy (86% of the renewable energy produced in 1993) and a low adoption of wind and geothermal energy technologies (respectively 1.8% and 2.7% of renewable energy produced) (see Table 4.3).

Table 4.3. Electric Power Installed in US (MW) - Absolute and percentage values

	1985	**1987**	**1989**	**1991**	**1993**
Sources					
Fossil	539921	548293	560524	575001	586129
Nuclear	87031	101604	106748	108443	107849
Others*	0.725	0.35	0.209	0.197	0.203
Renewable sources	91861	97861	103342	105977	108251
- Hydro	85558	87537	90094	91331	93205
- Geo	1967	2081	2681	2663	2944
- Biomass	2965	6383	8488	9827	10109
- Solar	0.096	0.193	0.329	0.392	0.393
- Wind	1371	1860	2079	2156	1993
Total	**718813**	**747758**	**770614**	**789421**	**802229**

Source: Energy Information Administration of the Department of Energy (US), quoted in Corvi (1996)

	1985	**1987**	**1989**	**1991**	**1993**
Sources					
Fossil	75.11	73.32	72.74	72.84	73.06
Nuclear	12.11	13.59	13.85	13.74	13.44
Others*	0.00	0.00	0.00	0.00	0.00
Renewable Sources	12.78	13.09	13.41	13.42	13.49
- Hydro	93.14	89.45	87.18	86.18	86.10
- Geo	2.14	2.13	2.59	2.51	2.72
- Biomass	3.23	6.52	8.21	9.27	9.34
- Solar	0.00	0.00	0.00	0.00	0.00
- Wind	1.49	1.90	2.01	2.03	1.84
Total	**100**	**100**	**100**	**100**	**100**

* encompasses agricultural waste, chemical processes, biogas, hydrogen

A different penetration level exists among various countries. Table 4.4 witnesses a large discrepancy in wind energy produced in different countries: in 1994, the US had the highest share of installed MW (nearly 50% of the world wind energy capacity), while for European countries, the share is rather low, with the exception of Germany and Denmark representing, respectively, 18% and 15% of the capacity installed in the world.

Table 4.4. Wind Energy Installed in the World - 1994

Countries	MW	%
US	1717	49.26
Other American Countries	9	0.26
Belgium	7	0.20
Denmark	539	15.46
Czech Republic	3.6	0.10
Finland	4	0.11
France	4	0.11
Germany	632.2	18.14
Greece	35.8	1.03
Italy	22	0.63
Ireland	8	0.23
The Netherlands	162	4.65
Norway	4	0.11
Portugal	8.5	0.24
Spain	72.6	2.08
Sweden	40	1.15
United Kingdom	170.5	4.89
Rest of Europe	9	0.26
Africa and Middle East	37.5	1.08
World	**3485.7**	**100**
- of which Europe	1722.2	

Source: European Wind Energy Association (EWEA), quoted in Corvi (1996); Bressan (1996)

One of the most common explanations for the low penetration of these technologies is the adverse climatic conditions, which do not guarantee the presence of the primary energy source necessary to develop renewable technologies. A simple interpretation of why photovoltaic systems are very limited in Northern European countries is often found in the limited number of hours of sunshine. Although this obvious remark may help in understanding the unsuccessful penetration of single technologies in particular countries, it is certainly not the only explanation for the low penetration achieved. Great Britain is an example in this respect: although in this country the Southern and the Scottish coasts have sufficient average wind force (respectively 5 m/s and 7 m/s) to host wind energy plants, the share of energy produced with wind plants amounts only to 4.8%, lower than in Germany and Denmark (Bressan 1996).

The data on the low penetration level of renewable forms of energy underline that a paradox exists between on the one hand the emphasis given to renewable technologies in the process towards a sustainable urban economy, and on the other hand the remarkably low penetration rate. This paradox emphasizes the need for initiatives in terms of clear policy strategies to stimulate a widespread diffusion of these technologies, especially at the urban level, but it also underlines the need for a conceptual framework that is able to identify barriers and bottlenecks in the adoption of these technologies.

In this chapter the aim is to present a conceptual economic framework on which our interpretation of the above mentioned paradox is based. According to our approach, the main reasons for the difficulties in the introduction of new technologies lie in the *techno-economic features of renewable technologies* which influence the adoption costs and revenues, and the way they evolve over time. In fact, our approach is based on the assumption that, since the techno-economic features characterising renewable technologies are different from those of conventional energy systems, the adoption of renewable energy technologies implies a radical change from previous technologies. It implies a break in the internal (micro-economic) learning processes of suppliers (learning by doing, by using, by searching, and indirectly, learning to learn), and of final users (new life styles, new social behaviours), as well as in the macro-institutional learning at the aggregate societal level. Technological change implies the generation of a new technological and organizational trajectory, based on new competence at both the level of suppliers (the energy producers) and final users[2].

---

[2] In this sector, adopters are conceived of as both energy suppliers and final users.

In the words of the evolutionary theory on innovation diffusion[3], all this means that the introduction of renewable energy technologies implies a change in the *technological paradigm* governing the economy and the society and it is in this perspective that the study of its adoption and of the bottlenecks and obstacles governing its introduction is undertaken. According to this theory, a new technological paradigm is defined as: "*a 'pattern' of solution of selected techno-economic problems based on highly selected principles derived from the natural sciences, jointly with specific rules aimed to acquire new knowledge and safeguard it, whenever possible, against rapid diffusion to the competitors*" (Dosi 1988, pp. 1127). Therefore, the study of a technological paradigmatic shift requires a dynamic approach. Time has to be seen as the rhythm of innovation, the pace of cumulative learning processes generating irreversible patterns and choices, where a more efficient technology may never become profitable since adopters are locked into the cumulative competence processes and in the positive feed-back effects stemming from the old technology (Arthur 1988 and 1990). This concept of time is the one applied in Part B of this study; and it is in this perspective that the analysis of the optimal adoption time is carried out.

This perspective is developed under different institutional regimes. As underlined in Section 3.4, the institutional changes taking place at the national level in most European Countries in the energy sector radically influence the rhythm of innovation, the strategies of local and national actors and the learning processes associated with the introduction of the new technology. For this reason, the conceptual framework is developed under different institutional hypotheses. This is the subject matter of Chapters 5 and 6, respectively, on monopoly and full competition regimes. In particular, in these two chapters the aim of our conceptual model is presented, namely:

- to analyse costs and revenues of adoption of renewable technologies in the perspective of a change towards a new paradigm. Different categories of costs and revenues are conceptualised, and their time path analysed;
- to define the optimal adoption time when particular techno-economic features characterize the new technologies. As we will demonstrate in our analysis, the profound changes towards a new technological paradigm gives rise to irreversibilities in the decision-making processes of adoption, which risk hampering the development of the new technologies over a long period;

[3] See, among others, Dosi (1982), Dosi (1988), Dosi *et al.* (1988), Freeman (1987), Nelson and Winter (1977; 1982).

- to propose a new role that the local public administration may play in the adoption process, in terms of intervention policies which may be more suitable to support the adoption process. The impression is that in view of a change in the technological paradigm, even the traditional (neoclassical) price-change strategy of different energy sources may turn out to be insufficient to encourage the adoption process in a city.

The first step towards the achievement of these goals is the identification of the techno-economic features of renewable technologies. This is the subject matter of the next section.

4.2 Techno-economic Features of Renewable Energy Technologies

As mentioned before, renewable energies encompass a wide spectrum of options ranging from solar, thermal or wind energy to photovoltaic energy systems and combined heat and power (CHP) applications. From a technical point of view, the elements on which an urban energy strategy can be put in place, among which also renewable energies can be mentioned, are threefold (OECD 1995):

- *energy efficiency improvements*, which increase the productivity of primary energy sources. It is the case of electric appliances, which are built with energy saving technologies;
- *combined heat and power generation*, which allows a more efficient conversion of primary sources into usable energy forms (CHP);
- *renewable energy sources*; i.e. those energy sources which may be reproduced, and therefore are not subject to depletion.

All these cases represent a way of introducing renewable energy technologies in an energy production system. An important difference characterizes the first case of renewable energy processes with respect to the others mentioned above. Its peculiarity stems from the fact that energy efficiency improvements are achieved through incremental innovations, *à la* Freeman (1987), on the already existing technologies, where the technological changes do not require any disruptive shifts towards a new technological paradigm, but follow the already existing technological trajectory, by increasing the efficiency of the already existing technologies.

The adoption process of these incremental innovations may be influenced by the relative prices of the technologies. The price of energy input factors related to the production of energy is in fact the most important variable in the decision making process of adoption, when no other costs are involved in the change other than the simple cost of purchase of the new technology.

A different adoption mechanism is behind the diffusion process of new CHP generation, and the use of renewable energy source technologies, like wind and solar electric power systems. The complexity and systemic nature of new technologies are extremely important in the definition of time and the penetration level of renewable technologies. Since they are based on new learning processes, new competences and new organizational solutions, at both the demand and supply side, they require a break with previous technological trajectories, previous know-how and organizational solutions, and thus imply high adoption costs.

The techno-economic features which characterize the new technological trajectory and its development may be easily classified in a matrix, according to their characteristics (technological, organizational, financial, economic, and environmental) and their economic aspects. The latter may be labelled as: indivisibilities, complementary assets, adjustment costs and sunk costs (Table 4.5).

The aim of the section is to highlight briefly these techno-economic features which influence the costs of adoption of renewable energy technologies.

4.2.1 Indivisibilities

Technological indivisibilities are faced when production plants cannot be built in small sizes, without incurring diseconomies of scale due to a sub-optimal size. Renewable energy plants are characterized by technological indivisibilities, as witnessed by diseconomies of scale existing in the case of small size plants: the cost of a particular type of solar thermal and power plant (WIS) is 2,460 $/kW for a plant with a size of 34 MWe, while in the case of a 600 kWe plant the specific plant cost has been estimated of 3,825 $/kW (Yogev and Kribus 1996). Technical economies of scale and learning processes are present also in photovoltaic plants and ocean thermal energy convertion technologies.

Table 4.5. The Techno-economic Features of Renewable Technologies

	Technological	Organizational	Market	Financial	Ecological/ Institutional
Indivisibility	Large size plants at start		High R&D activities Critical mass of users	High investment costs at the beginning Long pay-back periods	Nation-wide campaigns for public awareness of ecological problems
Complementary assets	Interface for distribution network security New heat networks		Know-how in new complementary technologies		
Adjustment costs	New and complementary grid networks Retrofit costs for private and public buildings	Know-how for the management of network grids open to private energy suppliers			
Sunk costs		Incomplete depreciation of previous energy production plants			

As Figure 4.1 shows, cost advantages are generated by both the size of the plant and the quantity of units of energy produced (Lindley 1996; Sjunnesson 1996).

Because of technological indivisibilities, many renewable energy technologies require substantial additional development or market penetration before they can produce under efficiency goals and make significant contributions to national energy supplies. Because most of these technologies are capital intensive, *the achievement of a mass production* (i.e. an urban, or better nationally widespread adoption) is essential to exploit economies of scale in the use of the new capital stocks.

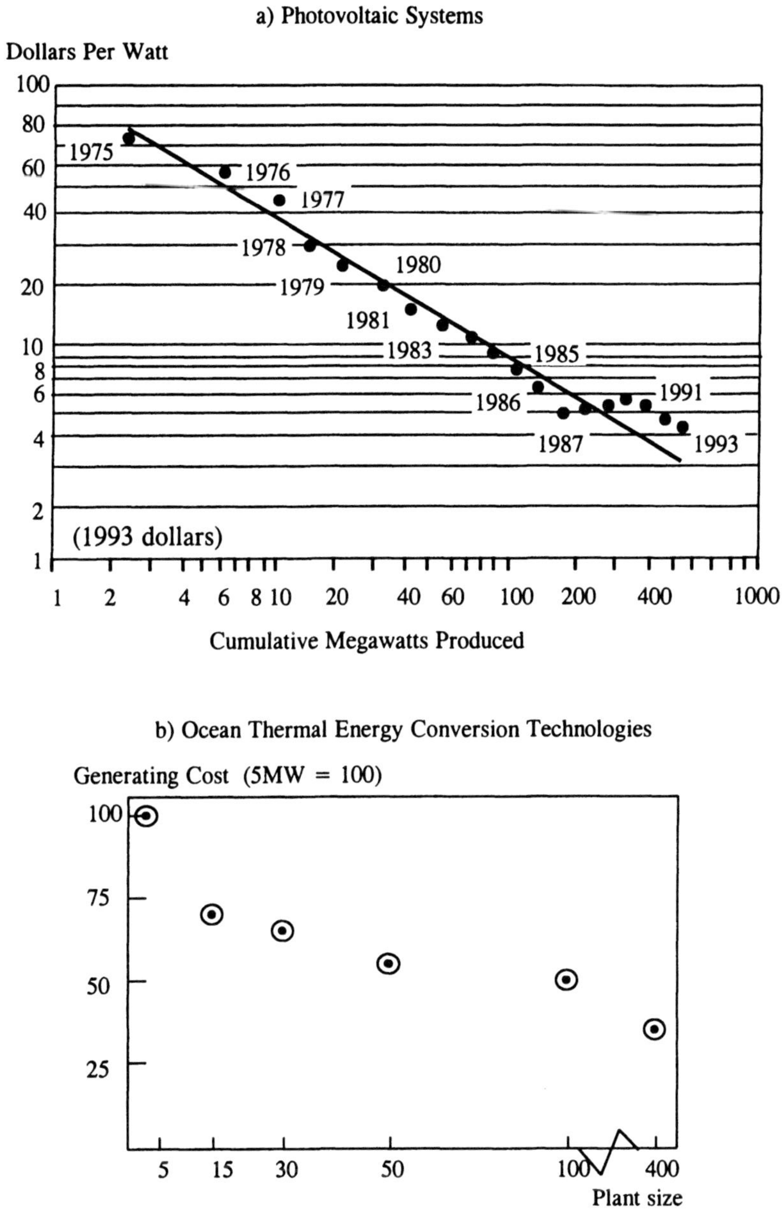

Source: IEA (1987)

Figure 4.1. Learning Processes in Photovoltaics and Economics of Scale in OTEC (Ocean Thermal Energy Conversion) Technologies

Mass production could in fact substantially lower the high fixed capital costs of some technologies, and thereby positively influence their market penetration. The renewable energy technologies which are most dependent on mass production for costs are at present active solar space and water heating systems, small to medium sized wind systems, parabolic troughs and small photovoltaics systems (IEA 1987).

Indivisibilities in renewable energy systems manifest themselves also at the economic level (Table 4.5). In view of technologies whose value very much depends on mass production to guarantee optimal production levels, the achievement of *a critical mass of adopters* becomes an essential aspect in order to generate a self-sustained adoption mechanism. This is a typical mechanism associated with the development of combined heat and power generation (CHP) in combination with District Heating (DH). Combined heat and power generation refers to the simultaneous production (cogeneration) and exploitation of high energy (mechanical and/or electrical) and low grade energy (thermal) from the same energy source. This system can be linked to a local district heating, thus providing CHP within a specific area. In this case, the profitability of the plant is linked to the number of potential adopters, who share the implementation costs of the new plant.

In early adoption stages, renewable technologies require *huge R&D efforts* before they can be economically competitive with conventional energy systems, especially with current electricity grid networks (IEA 1994b). R&D expenditures represent the critical success factor in leading these technologies towards commercialization.

A particular stock of scientific competence is necessary to supply energy in the case of the new systems. However, other kinds of know-how acquisitions have to be put in place. Commercialization campaigns of renewable technologies are rather different from those of traditional technologies. New technologies require an ad-hoc promotion for single final users (or groups of users). In the case of passive solar systems, ad-hoc solutions have to be proposed each time to solve problems of retrofitting to existing buildings and to keep sunk costs and financial losses as low as possible. These solutions imply and require marketing strategies to be tailored to individual users[4], while mass commercialisation campaigns are not appropriate in this respect, as they were for traditional technologies

---

[4] A similar case of new promotion strategies is in the field of new telecommunications technologies, which differ from the previous traditional telecommunications networks (Camagni and Capello 1991; Capello and Williams 1992).

characterized by relatively homogeneous product markets and mass produced outputs. Appropriate commercialisation campaigns determine positive experiences and success stories of adoption, which play an important role in stimulating and encouraging potential adopters. Paradoxically, up to now the public perception of CHP is so far based in most cases on previous bad experiences with poorly designed and constructed systems. Lack of consumer heating control, unreliable systems resulting from delays in repairs and maintenance, have given district heating (DH) a bad reputation in some countries (OECD 1995).

Another area where a new stock of knowledge is very important in order to help the development of new energy technologies, is the governmental level. In this area, radical changes in the urban planning system have to take place, when renewable energy supply technologies are introduced at the urban level. The most difficult and strategic change is the *integration* between different urban functions (Fusco Girard and Nijkamp 1997). The main strategic integration should take place among the following sectors:

- transport management, land use and energy supply;
- urban planning and energy supply.

In the first case, an example of the integration among transport management, land use and energy supply is the exploitation of transport biofuels. Special energy crops can be grown to produce substitute road transport fuels using currently proven agricultural practices and technology. For example, sugar cane and maize can be used to produce ethanol and for mixing with petrol to obtain gasohol (OECD 1995). The commercial production of these kinds of energy supply systems depends very much on some strategic aspects, namely: a) the co-operation of existing road transport fuel suppliers, especially in relation to blending the provision of alternative fuels with conventional fuels at filling stations, b) the amount of land required to grow special energy crops. Problems could easily arise from the conflict between land use for food and for energy production, when ex-ante integrated decisions are not made at the level of transport managers, energy providers and urban planners.

The second strategic integration when renewable energy supply systems are introduced at the urban level takes place simply between physical urban planners and energy supply strategies. An example of the strategic importance of this integration is passive solar power technology. Modern building designs, building practices and architectural features of new buildings should not be heavily dependent on artificial cooling by means of air conditioning equipment, and should have all technical characteristics required to implement renewable energy supply systems. In the same way, a quick connection of new districts in the city with CHP systems is made available when the physical planning of the new buildings takes into account the technical aspects required in order to provide energy with renewable technology systems in the new area. In general, urban planning should take into account siting, design and orientation of new residential areas, when for example passive solar energy should be introduced. In summary, an energy-conscious urban development policy can contribute to reducing both the intrinsic energy needs of building and the efficiency with which these needs are met.

Indivisibilities manifest themselves also in the necessity to put in place nationally widespread campaigns to support ecological awareness and culture, at the level of both firms and citizens. A positive attitude towards environmental sources would increase the value of the social advantages associated with energy saving and environmental protection. In fact, the willingness to pay for environmental protection highly depends on the cultural attitude of each person, which introduces in his utility function also social environmental goods (like air quality). It is the area where energy producers (or potential energy producers) merge their effort in a common alliance for an institutional recognition of the importance of these technologies for the environment.

The same kind of ecological awareness is important for local governments to achieve social consensus on some renewable energy technologies. For example, the introduction of a single or several wind machines has visual effects similar to power transmission pylons on the landscape. A solution like this in an urban peripheral area may face strong opposition by residents if an ecological culture is not widely developed. The same can be said for the introduction of passive solar technologies in architectural design which can affect building aesthetics. Such new architectural designs are more easily accepted by potential owners of passive buildings when an ecological culture is more widely spread out.

4.2.2 Complementary assets

Renewable energy plants require a set of complementary technologies for their implementation. Some of them are evident, like distribution grids, others are more common to technicians, like interfaces for the security of the distribution network. In this respect, renewable energy technologies are systemic technologies and a change in a technological trajectory of the core technology implies changes in all complementary assets.

Complementary assets require both radical as well as incremental changes when renewable technologies are introduced. An example of radical change in complementary assets is district heating (DH). For this new heating system, the implementation of a new hot water distribution grid is required. The same holds good for technological systems used to cumulate energy when this is produced by a discontinuous source, like wind and sun. These systems have to be implemented ex-novo, when new wind and photovoltaic power plants are installed. Incremental adjustment changes have to be borne in the area of control systems for the distribution grid. If energy is produced and inserted in the distribution network by more than one agent, sophisticated electronic systems used to manage and run the electric distribution network are necessary.

Especially for radical technological changes in complementary technologies, new competences have to be developed and training of new staff implemented. This is the case of new technical know-how for the maintenance of distribution grid when this is open to more than one energy supplier. In this respect, skills, knowledge and specialization of previous staff represent a quasi-irreversible process while time passes, since firms lock-in their competences and cumulate their know-how around the conventional technology.

4.2.3 Adjustment costs

The shift from one technological paradigm to another implies for potential adopters adjustment costs from the old to the new technology (Scherer 1980), costs which have in the past often been underestimated in the analysis of innovation diffusion (Camagni 1985). These costs are of both a technological as well as an organizational nature, and in the case of renewable technologies, they have to be borne by both energy suppliers already producing with conventional power plants and final users who are interested in the new technology.

In the case of energy suppliers, examples of adjustment costs are new distribution grids which have to be developed and managed, once new power plants are installed. Many large power stations are located remote from the centres for reasons of economy, environmental protection, fuel storage, and so on. The cost of new transmission lines can be prohibitive. Up to 1987, the maximum economically viable transmission distance of new district heating distribution networks was up to 30 km for hot water and 3-5 km for steam, depending on the heat load and fuel prices (IEA 1987). In many circumstances it may be more economic to construct new heat production units at short distances from the demand centres, than to construct a long transmission line. However, also the exploitation of conventional existing distribution networks for new technologies implies adjustment costs which hamper the widespread adoption of district heating.

As regards adjustment costs for final users, these are typically related to retrofit costs of already existing heating systems in houses, or in buildings. One of the most important factors in the development of renewable technologies from the final users side, especially new district heating systems, is whether they have been constructed as a new development or as a retrofit development. For the former, costs are much lower, while for a retrofit situation costs can be prohibitive and the degree of disruption can be significant (IEA, 1987). Energy producers bear the problem of retrofitting costs of conventional technologies condensing power plants for combined heat and power production. Technical solutions to the loss in power output involved are necessary, as well as adjustments to orient the heat output to be fed into the heating grid. Private groups of adopters meet high retrofitting costs at the level of existing buildings. The new energy supply systems cannot make use of the already existing heating plant, which has to be substituted with the new one. In most cases the introduction of such new heating plants requires changes in the building structure. Passive solar power technology involves incorporating features into the design and construction of buildings so that they tap the heat available for use in space heating and cooling. Since passive solar features are inherently a part of the design of a building, they are easier to incorporate into new buildings, with consequent high retrofit costs on the old building stocks (OECD 1995). In this case, the adjustment costs create barriers for faster introduction, since the rate at which passive solar technology is adopted is usually limited to the rate at which new buildings are constructed (Table 4.5).

Private adopters also face retrofit costs in the case of in-house heating systems. The cost of converting from a conventional hot water individual heating system to district heating can range from US$ 20 to US$ 200 per kW. For the conversion of in-house steam heating systems to hot water district heating the costs have been estimated to range from US$ 70 to US$ 400 per kW (IEA 1987). These costs must be borne by the consumer and can be a significant deterrent to district heating connections.

At the aggregate level, retrofit costs still exist and represent a disincentive to the introduction of renewable technology. Retrofit costs are present in the case of public buildings, whose old heating systems require the same costs of private buildings to be adjusted to the new technologies.

4.2.4 Sunk costs

In most cases, traditional plants supplying energy from fuel and fossil energy sources are incompatible with the supply of energy from new energy sources, like wind and sun. Sunk costs on incomplete depreciation of previous equipment have to be borne by an energy producer when new technology systems for supplying energy are adopted (Table 4.5).

Also in the case of the introduction of new energy supply systems for the whole city, sunk costs on incomplete depreciation of energy supply networks arise. The distribution of energy supplied by renewable sources via existing urban, regional and national electricity networks is possible in some cases, although with very high losses in terms of produced energy (Coiante 1992). In the absence of a dedicated network for renewable energy systems, however, one must focus on existing infrastructure which, unfortunately, tends to treat renewable energy sources as a sideline opportunity (e.g. plumbing industry for solar domestic water heaters, electric utilities for wind) with only limited success. The possibility of exploiting the existing electric network is not so simple from a technical point of view, does not represent the majority of techniques to which it can be linked, and is sometimes inefficient. In fact, even though it can be useful during the initial stages of market penetration, because it speeds up the availability of supplying energy with renewable systems, it is not an adequate strategy for full penetration, because of the technical inefficiencies which are generated. If this is the case, a long term full penetration strategy has inevitably to face huge sunk costs on an incomplete depreciation of previous energy supply networks.

4.3 Development Trajectories of Renewable Technologies

Renewable technologies play an important role in the control of urban environment and energy use if they are introduced in order to substitute traditional energy technologies in the production of a large share of energy requirements. Only in this case, the impact on negative environmental factors and costs of energy consumption could be partially solved and kept under control. However, to achieve the substitution phase, barriers and bottlenecks have to be removed and overcome, and the techno-economic features of these technologies need to be fully accepted by both suppliers and final users.

As the data presented in Section 4.1 illustrates, the situation is at present far from a substitution phase. Stand-alone technology plants are at present diffused, especially in remote areas, like small villages and remote islands where the costs of implementing an energy distribution network would be more expansive than installing a small wind energy plant, or a small PV system. Examples of this kind of application exist in each country, like for example small stand-alone PV systems in remote islands in Greece, and small stand-alone hydro power stations in mountain villages in Italy.

The stand alone systems, however, do not guarantee the widespread diffusion of these technologies. The diffusion processes of renewable technologies may be summarized in three distinct phases, which may all have a traditional logistic or S-shaped curve, as shown in Figure 4.2.

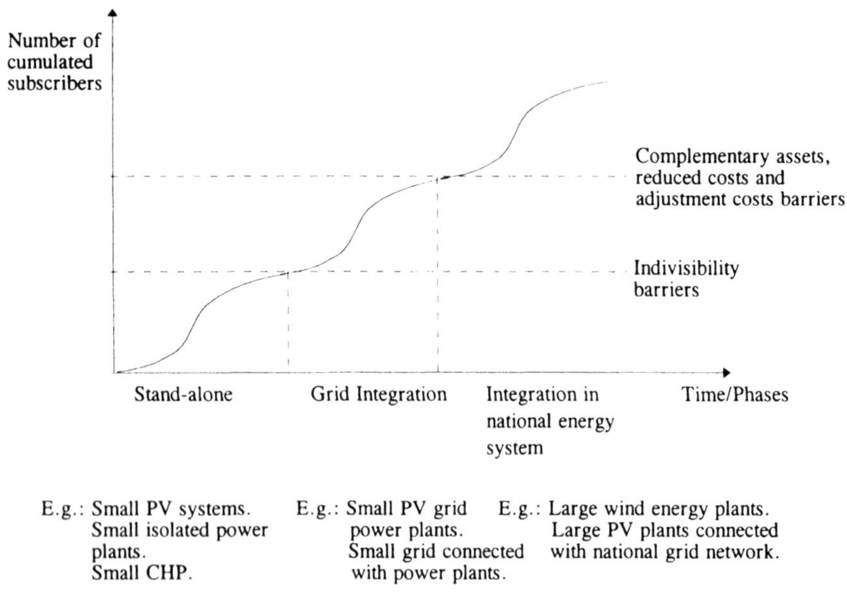

Figure 4.2. The Three Adoption Phases of Renewable Technologies

These phases correspond to different integration strategies within the national energy system, moving from a stand-alone adoption approach, to a full integration, with the aim of substituting conventional plants for renewable energy technologies. The three different phases may be labelled as:

- the *stand-alone phase*, in which renewable energy plants are not integrated in the national energy system, in terms of distribution network and energy supply. This is the most simple phase to reach, since it requires a simple *relative profitability strategy* in the decision-making process of adoption. The profitability of renewable plants, compared with the profitability of enlarging the distribution network and the energy production capacity of traditional plants, reflects clearly a more profitable solution in favour of remote, isolated and, consequently low demand areas. In this phase the number of users who may adopt these technologies is limited and thus the adoption tends towards a saturation level for very low levels of cumulated adopters. To overcome such a saturation level, a different adoption strategy is required, viz. a strategy which allows new plants to integrate into the national distribution networks, giving rise to a second phase of development;
- the *network integration phase*, in which renewable technologies overcome the indivisibility barriers, by achieving a critical mass of users and thus by exploiting decreasing average costs. In this phase, the minimum threshold size of plants is reached, and thus technological and market indivisibilities are overcome. This is the case of district heating systems connected with the distribution grid in small areas of the city, where a sufficient number of subscribers is linked to the new production plant and the adoption costs are thus divided among a sufficient number of people. However, small power plants integrated with a distribution grid are not sufficient to produce enough energy to substitute conventional plants, but at the same time they do not require to overcome the barriers of sunk and adjustment costs, as in the case of the large plants integrated in the nationally wide network, typical of the third phase;
- the *large size plants integration phase*, in which renewable technologies substitute a high share of energy produced via conventional technology plants. At this level, all adjustment and sunk costs associated with the transition phase from the old to the new paradigm have to be overcome, and renewable technologies become a real substitute for traditional fossil fuel systems. At present, given the costs of implementation of large renewable energy plants, the relatively low demand and the sunk costs associated with the low

depreciation of existing plants, the profitability of the new and large renewable plants is too high. However, as we will argue in the next chapters, an approach giving emphasis to a dynamic process, where adoption costs and advantages are estimated as the discount value of future adoption costs and advantages may demonstrate that, even though at present these technologies may have a low profitability value, a strategy of postponing the adoption may even lead to a situation when these technologies are no longer profitable.

Table 4.6 provides the different features characterizing the three phases of development. As mentioned above, the stand-alone phase is characterized by small and isolated plants, which are destined to a niche market. The barriers to the adoption of these kinds of technologies are nearly non-existent. The core technology is at a mature stage, while still a place for technological improvements exists for complementary assets.

For the integration phase, the small grid connected plants are already a reality from the technological point of view: no technological barrier exists for their implementation. More problems and bottlenecks in the transition to this phase can be found in the size of plants, which requires a minimum size to be profitable and to exploit economies of scale. This also means that these kinds of technologies imply a critical mass of users, and a certain market size. Barriers exist in this phase at the organizational level: the energy distribution network requires a particular know-how in order to manage the distribution network exploited by different producers. In other words, one can summarize the bottlenecks to achieve this phase as market barriers (Table 4.6).

The last phase is characterized by the implementation of large plants, connected to the national distribution systems, which substitute a high percentage of energy produced with traditional plants. These technologies refer to a national market, and for this reason are widely-spread. At present, this phase is far from representing the real situation, as witnessed by the data on the share of renewable energy produced by firms, since the achievement of a widespread development of these technologies requires to overcome technological, organizational and market barriers.

Table 4.6. A Comparison of the Three Phases of Development

Phases / **Elements**	**Stand-alone**	**Grid network integration of small energy systems**	**Grid network integration of large energy systems**
Technology	Small isolated power plants	Small grid connected power plants	Large grid connected power plants
Market features	Niches	Small scale market	Nation-wide market
Entry barriers to the phase			
* technical	None. Only for particular complementary assets	Minimum size of plants	Complementary assets for grid connection
* organizational	None	Managerial know-how of the grid network	Functional know-how of the grid network
* market	None	Critical mass of users	Competition with traditional technologies
Degree of maturity of the present technology	Core technology: high Complementary assets: low	Core technology: high Complementary assets: low	Core technology: low Complementary assets: low
Barriers to development	None	Market	Technological and market

As Table 4.6 shows, the technological problems in this phase arise from the non-mature core and complementary assets technologies. Technological problems of this kind relate to two aspects:

- the grid / plant interface. The stability of the energy intensity on the grid network, in order to avoid on unexpected surplus or deficit of energy in any moment in time, requires a control of the energy produced put in place by new grid/plant interfaces and by new management procedures;
- the software systems to manage the grid. Any technical maintenance on the grid requires total security measures, i.e. the total absence of energy at the time of the intervention. The security measures are guaranteed by software systems which interrupt the supply of energy on a particular part of the grid. Sophisticated software systems, which still require technological upgrading, have to be put in place when there are multiple sources of energy supply.

The third and more strategic phase has to be achieved, if renewable technologies aim to play a role in improving environmental quality and in keeping negative factors under control. A serious effort has then to be made to overcome the above mentioned barriers, and to achieve a widespread adoption of these (potentially) rather important technologies.

4.4 Conclusions

The importance of renewable energy systems as an efficient policy option to keep environmental costs under control is widely accepted among experts and policy-makers; however, in analysing the data on the penetration level of these new technologies, a striking result emerges on the very low penetration level of these technologies. This is true for a wide range of possible technological solutions, from CHP to wind and photovoltaic systems. This low adoption trends characterizes the US and most European countries, with some exceptions related to particular technologies.

The low penetration level is in general explained by simple arguments, like adverse climatic conditions which do not guarantee the presence of the primary energy source, too a few hours of sunshine per day, or low wind force. Our approach to the existing low penetration level is somehow different, and draws the attention to the techno-economic features characterizing these new technologies. The difference in the learning processes required by the new technologies are so different from the previous conventional technologies, that one may plausibly argue that a new techno-economic paradigm is going to take place. However, this change of paradigm is in our view the most crucial element in the definition of the rhythm of adoption and diffusion of new technologies, and may explain the low penetration level.

In this chapter, a specific taxonomy of techno-economic features has been presented. These features turn into adoption costs in our model on the optimal adoption time, since they represent entry barriers to the new paradigm, as well as exit barriers from the old one. These barriers weigh differently, according to the phase of development in which the technologies are: the stand-alone phase is not required to overcome any particular barrier for its development, since it does not achieve any substitution with the more conventional energy systems. However, both the network integration phase and the large size plants integration phase need to overcome the techno-economic barriers characterizing the move from one paradigm to the other.

Our aim in part B of this study is to build a theoretical model of optimal adoption time when the adoption implies a change in the technological paradigm. The model will be based on the comparison between adoption costs and revenues, which evolve over time according to the institutional rules governing the industry.

5 Irreversibilities and Indivisibilities of Renewable Energy Adoption Processes in a Monopolistic Market

5.1 Introduction[1]

In this chapter we present a conceptual model of adoption processes under the assumption that renewable energy is supplied in a monopolistic market. As anticipated in Chapter 4, our approach to the adoption processes and to the choice of the optimal moment of adoption is a dynamic approach, since it stresses the way in which adoption costs and revenues change while time passes. Learning processes and cumulative know-how on the old technology influence the values of adoption costs and advantages over time, and may in the long run turn a profitable adoption into a non profitable solution.

Our approach is based on the comparison of the discounted values of adoption costs and revenue over time. A situation is profitable when revenue exceeds cost, and the optimal time of adoption is when the difference between revenue and cost is at a maximum. The model builds a typology of costs and revenues, presented in Section 5.2, and especially analyses the development of their discounted values over time. The typology of costs presented in the next section stems from the structural features characterising renewable technologies which have been stressed in Chapter 4, as they represent the different categories of costs to be borne for the transition from an old to a new paradigm.

[1] This chapter is written on the basis of previous work developed by Camagni and Capello (1998).

Two important hypotheses characterize the theoretical model. The first hypothesis is that the renewable energy is supplied in a monopolistic market, and this hypothesis explains, as we will see, the constant value of adoption revenues over time. The second hypothesis, which is a consequence of the first one, is that the firm which innovates and enters the renewable energy market is the first innovator, whenever he decides to enter. This hypothesis explains the possibility in the model to have profitable adoption solutions which do not lead to innovation; the potential innovator is conscious that he can wait to enter a profitable market, since nobody will take his place.

Moreover, as we will see, the model is developed for both a potential innovator who operates already in the conventional energy market and may thus decide to enter the renewable energy market by jumping from one paradigm to the other, and for agents who have never operated in the energy market. As we will see, the optimal time of adoption, and thus the innovation behaviour, is different.

5.2 Adoption Costs and Revenues of the New Technological Paradigm: a Dynamic Perspective

According to the evolutionary school of innovation economics (Dosi 1982; Dosi *et al.*, 1988, Nelson and Winter 1977, 1982), technological change may be interpreted as an irreversible, path-dependent and evolutionary process, stemming from the behaviour of economic agents who explore only a limited part of the set of theoretically possible actions, viz. that part that is strictly linked to previous innovation adoption and to already acquired know-how. Its path-dependent nature and the non-linearities connected with the learning processes may act as dynamic *entry barriers* with respect to possible, perhaps more efficient, alternative technologies. Existing adopters of consolidated technologies benefit from positive feedback effects, increasingly consolidating the advantage of these technologies and widening the adoption costs of the new technology. Dynamic barriers for new entrants work at the same time as *exit barriers* for existing actors, perpetuating the actual market shares of old and new technologies well beyond the situation expected on the basis of their relative efficiency of the core technologies[2] (David 1985).

All this holds good for each technological trajectory: a change in the technological paradigm requires a break with the previous technology, a break with the already existing learning processes, and the starting of new patterns of knowledge. For its cumulative and path-dependent nature, the development of a new paradigm is not guaranteed by the technical superiority of the incorporated technology. It is also necessary that the discounted value of the differential earnings expected from the new technology be higher than the costs which have to be faced to bring the new technological trajectory into line (Camagni and Cappellin, 1984). In this perspective, the discounted values of the different categories of costs and advantages over time allows to highlight the time in which the supply is profitable from an economic point of view.

Table 5.1 presents a taxonomy of supply costs and revenues and their trend over time. The supply costs may be classified into:

- *entry barriers*, which reflect the supply costs associated with the decision of every economic agent who is willing to innovate and thus enter the new technological paradigm;
- *exit barriers*, which reflect the adoption costs related to the change from the old to the new paradigm, for those economic agents who have to leave a previous technological trajectory and enter the new one.

[2] The example presented by David (1985) on the Qwerty keyboard for typewriters is illustrative for this sort of "market failure". Although the Dvorak keyboard was vastly superior to Qwerty, virtually it has never been widely adopted because of a) *technical interrelatedness*, i.e. the need for system compatibility between hardware (the machine) and software (the typists' memory of a particular arrangement of the keys) which makes the value of the hardware highly dependent on the existence of the compatible software; b) *scale economies*, i.e. decreasing cost conditions while increasing the number of people using a particular system; c) *quasi-irreversibility* of investment in specific touch-typing skills (David 1985 and 1992; Liebowitz and Margolis 1990).

Table 5.1. Taxonomy of supply costs and revenues, and their time trend

Costs of supply of the new technology	Description	Time trend (postponing supply)	Reasons for trend
Entry barriers			
a) Indivisibilities			
Technological	Large dimension at start	constant	
Market	Critical mass of customers	constant	
Organisational	Large and complex organisation at start	constant	
Competence	Wide stock of new competence	decreasing	context learning
Institutional	Necessary alliances with others	decreasing	context awareness
b) Complementary assets			
Technological	Change in customers' technology	increasing	mechanism of locking-in the old technology
Organisational	Change in customers' organisation	increasing	mechanism of locking-in the old technology
Competence	Complementary competence * of customers * of suppliers	increasing	mechanism of locking-in the old technology
Exit barriers			
a) Adjustment costs			
Organisational/ competence	Costs of organisational adjustments	constant and high	
b) Sunk costs			
Technological	Incomplete depreciation of capital stock	decreasing	Cumulative depreciation of existing capital stock

Table 5.1 (continued).

Revenues of supply of the new technology	Description	Time trend	Reasons for trend
Economic	Net revenues for internal suppliers	decreasing	Learning on the old technology
	Absolute revenues for external suppliers	constant	
	Exploitation of advantages of first adopters: * at the social level: curiosity, political approval * at the private level: product innovation	constant	

5.2.1 Adoption costs: entry barriers

The first category of costs envisaged, i.e. entry barriers, may be defined around two main categories: (i) indivisibilities and (ii) complementary assets.

(i) *Indivisibilities*
The shift towards a new technological paradigm requires large investments in the new technological trajectory, in order to build a certain productive capacity (Gaffard 1990). These investments have the characteristic to be indivisible since they have to be faced at the moment of adoption, and cannot be distributed over time. They can be of a different nature. They can be technological investments if they refer to the large size of the production plants required at the start in order to achieve economies of scale in the production processes. A similar condition may be required in the organizational sphere: a new and complex organization crucial for the new production. At the competence level, a change in the technological trajectory implies a new and large stock of competence. The old knowledge is by definition related to a different technological paradigm: the new trajectory is fed by a different scientific and technological competence, which is required from the beginning to start exploiting the new technologies. It is on this new competence that the development of new learning processes is based, and new cumulative patterns of development are put in place. Moreover, from an *institutional* point of view, during the adoption process the new technological trajectories require lobby

activity by potential suppliers to generate a cumulative agreement in the society on the advantages that the technological adoption may generate.

(ii) *Complementary assets*
The shift towards a new technological paradigm implies the reorganization of the development pattern around a radical innovation. However, around the core technology which drives the change, a number of interrelated technological upgradings in all complementary assets are needed. This is even more true when technological paradigms are characterized by an increasingly complex and systemic nature: in these cases, the radical innovation on which the new paradigm rests, influences the spectrum of technological complementary assets (Antonelli 1995). The same pervasive and drastic change is required in the organizational structure of suppliers and customers: they have to break the traditional organizational routines and to build new ones around new technological processes. Needless to say, the new technological paradigm imposes a new learning path even in all complementary assets of the core technology.

5.2.2 Adoption costs: exit barriers

The second category of adoption costs which may be foreseen is represented by *exit barriers*. As mentioned before, these costs regard the already existing producers in the market, who choose to move from the old to the new technology.

The intrinsic learning value of technological change generates exit barriers in the previous technology, since cumulative processes reinforce and perpetuate the advantages of that technology, highly reducing the possible interest to any alternative technology. These exit barriers are defined as adjustment costs and sunk costs.

(i) *Adjustment costs*
Adjustment costs are the counterpart of the high organizational indivisibilities that have to be faced by new entrants, when they decide to innovate. Adjustment costs regard in fact the organizational sphere of the firm which in the case of already present economic agents, does not need to be fully structured ex-novo, but requires an adjustment to the new paradigm. Sometimes, however, even the reorganization of labour and functions is expensive: in the perspective of the evolutionary school, this element is one of the main obstacles to the introduction of a new paradigm. Organizational routines may lock the firm in the previous

technological trajectory and limit both its willingness and interest to any other techno-organizational solution (Nelson and Winter 1977).

(ii) *Sunk costs*

The jump from one technological trajectory to another is also hampered by sunk costs on the previous technology, such as the incomplete depreciation of the previous capital stock. The full exploitation of the previous technology is an objective that the firm can legitimately hold. It also inhibits, however, the move towards the new technology in the short run. These elements may justify a conservative strategy, but by postponing the decision to innovate, they postpone over time also the starting of the new learning processes.

5.2.3 Adoption revenues

The counterpart of all these costs are the *revenues* that the introduction of a new technology generates. These advantages are of an economic nature. The adoption choice model we present hereafter analyses the adoption behaviour of energy suppliers, leaving aside the adoption behaviour of final users. For this reason, adoption advantages are in our perspective adoption revenues achieved by potential energy suppliers through renewable technology adoption. Under the assumption of a monopolistic market structure, revenues remain constant over time, since prices of energy also remain stable over time.

Another category of advantages stems from the condition of being the first supplier to introduce such a product innovation. This position would provide important returns as it would allow to exploit general psychological elements, such as curiosity and excitement about the novelty. These kinds of advantages hold good for both already existing energy suppliers and for new ones. These advantages remain constant over time, when the perspective of the first adopter is taken for granted. These pshycological positive elements that may help the adoption process strictly depend on the level of diffusion of the innovation; the first adopter will be able to exploit them, whenever the decision to innovate is made. For this reason, the advantage curve for the external suppliers is constant over time.

The choice of adoption is made in our model on the basis of a comparison between adoption costs and revenues: when the latter exceed the former, the adoption is profitable, since it assumes positive values. The optimal adoption time takes place when the distance between revenues and costs is maximum; at that moment, profitability assumes maximum value.

In our adoption choice model we will distinguish two different cases. In the first case, potential adopters are already existing producers in the conventional energy market (from now on "internal producers"). For them, the adoption choice is based on relative profitability: in fact, in this case, the absolute value of the profitability of the new technology is discounted by the opportunity cost to remain in the old paradigm, and supply energy with conventional technologies. This opportunity cost increases over time, because of learning processes which favour, through cumulative know-how mechanisms, to remain on the old technological trajectory. In the second case, on the contrary, adopters enter the energy market for the first time, and thus decide whether to adopt or not on the basis of absolute revenues.

5.3 The Optimal Time of Adoption: the Role of Irreversibilities

The dynamic approach to the adoption/supply of renewable energy technologies underlines the obstacles and problems associated with a move towards a new technological paradigm. However, this approach is also extremely useful to highlight the optimal time of adoption in the presence of learning processes, irreversible investments on previous technologies, and indivisibilities of new scientific, organizational, financial and technological investments. The situation differs between "external suppliers" and "internal ones".

The case of "external suppliers"

Let us start with the case of "external suppliers": Figure 5.1 presents the adoption revenue and cost curves over time, where revenues for "external suppliers" remain constant over time.

The adoption cost curve, on the contrary, shows an increasing slope over time, as a result of the sum of three different trends: a) a constant cost curve related to the value of the indivisibility costs which remain stable; b) an increasing cost curve, reflecting the growing size, complexity and efficiency of the complementary assets organized around the old technology, which should be replaced if the new technology is to be sold on the market; and c) a decreasing cost curve, due to context learning processes about the new technology which decrease the value of costs over time.

It is plausible to assume that context learning processes will decrease costs in the long run, but this positive impact will be very limited. On the contrary, the lock-in mechanisms associated with the old technological trajectory will have strong effects also in the short run, since they exploit a more declining learning process curve. If this is the case, the total adoption cost curve assumes an increasing slope.

Figure 5.1 summarizes the total cost and advantage curves obtained: while time passes, the advantages of adoption decrease, while costs increase. The logical consequence is that, if an optimal time of adoption exists, this is at time 0.

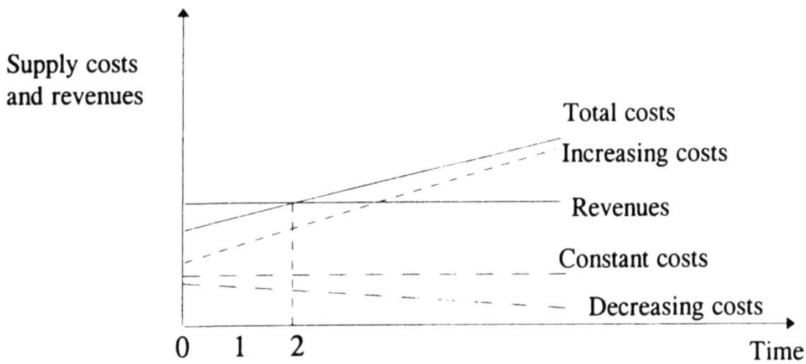

Figure 5.1. Supply costs and revenues of "external suppliers"

The result is that an early adoption is a more profitable choice; it allows the start of new learning processes on the new technological trajectory in a more recent time, and it especially helps avoid the growth of complementary assets and end users equipment around the old technology. At time 2 in Figure 5.1, the growth of these latter elements is so high as to overcome the revenues associated with the new technology: after time 2, the new technology is no longer profitable.

The case of "internal suppliers"

A different situation is the one related to already existing suppliers in the energy market. The total advantage and cost curves are presented in Figure 5.2. While the net revenue curve[3] decreases over time, the total cost curve changes drastically. "Internal suppliers" have to face exit barriers from the previous technological paradigm, if they want to innovate; the exit barriers decrease over time because they depend on the incomplete depreciation of the capital stock. It is reasonable to expect that the sunk costs on the incomplete depreciation of the already existing capital stock influence the total costs over time more than the mechanisms of continous learning on the old technology. If this is the case, the total cost curve assumes a negative slope until the previous capital stock is completely depreciated (at time 2 of Figure 5.2c). At that time, the adoption costs of the new technology start to increase, influenced by the learning processes around the old technological trajectory.

Figure 5.2 summarizes these trends; the results in terms of optimal time of adoption differ from the case of "external suppliers". For "internal suppliers", the optimal adoption time depends on the slope assumed by adoption costs and revenues. There are three possible situations:

a) the time derivative of the costs is lower than the one of the revenues. In this case (Figure 5.2a), the optimal point of adoption is time 0, as in the case of "external suppliers";
b) the time derivative of the costs is equal to the one of the adoption revenues. In this case (Figure 5.2b), the optimal time of adoption is in the range between time 0 and 2. Within this time span, the choice is optimal, though the technology is still profitable until time 3;
c) the time derivative of the costs exceeds the one of the revenues. In this case, for the already existing energy producers, the optimal time of adoption is at the point when the old capital stock has been completely depreciated (Figure 5.2c). If an optimal time of adoption ever exists, this is at time 2, when the difference between adoption costs and revenues is wider. The early adoption does not seem to be the right strategy for already existing energy suppliers: the sunk costs associated with their previous technology oblige them to wait until the depreciation process has ended.

---

[3] The net revenue curve represents the absolute revenue of the new technology adoption from which the opportunity costs of remaining on the old technological trajectories are subtracted.

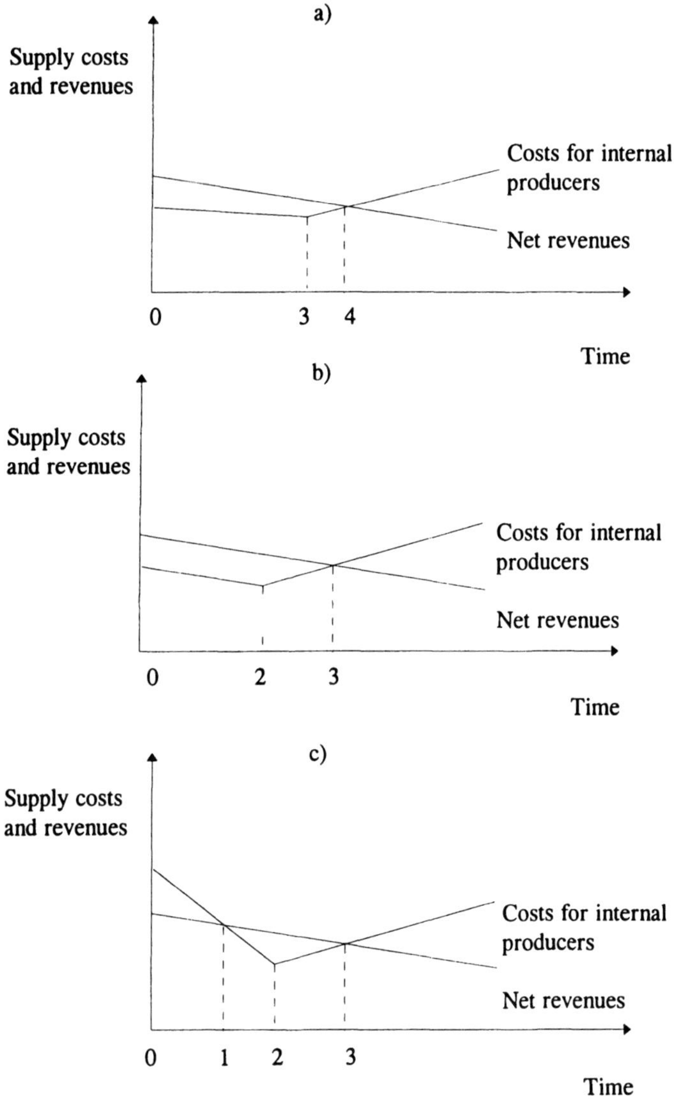

Figure 5.2. Supply costs and advantages for "internal suppliers"

Also for internal adopters, irreversibilities in the choice of adoption may emerge. As Figure 5.2 shows, in each of the three cases, if the choice of adoption is not made in a certain range of time, the technology will no longer be profitable.

Some interesting conclusions can be drawn from this part of our analysis. In the presence of irreversible lock-in mechanisms in the complementary assets of the new technology and of indivisibilities in the investments it requires, an early adoption is more profitable. The choice to wait implies paying higher entry costs in the future due to the irreversible investments associated with the old technology, with the growth of scientific competence and with the establishement of organizational routines within the old technologies.

The choice of a late adoption is a risky one also for already existing energy suppliers; for them, the best opportunity for adoption may be either immediately or after all sunk costs have become very low. However, very late adoption for them is also not profitable. The learning processes on the old technology may be so strong that in the long run they may perpetuate and even reinforce that choice, although they face the presence of a more efficient and less environmentally damaging technology.

The existence of irreversible investments on the old technology leads to *irreversible non-choices of adoption*: either the new technological trajectory is adopted in an early time span, or the new technological paradigm will never come into existence. In this case we are facing a different kind of irreversibility: the traditional irreversibility envisaged by the evolutionary theory, stems from the existence of sunk costs *and thus its cost is related to the old technology*. In the present case, on the contrary, irreversibility regards a crystallization of the decision of non-adoption and, under this perspective, *the cost of the irreversibility provokes its negative effects on the new technological paradigm*, since it impedes its development.

Our conceptual scheme does not guarantee, however, that the adoption of the new technology is, under all circumstances, profitable. It may also be the case that the new technology is never profitable, since costs always exceed revenues of adoption (Figure 5.3). If CT is the total cost curve, and A the advantage curve, the situation shows no profitable solutions: intervention policies are required in order to either increase the advantages or decrease the adoption costs.

In this context, another consideration stems from our analysis: in the presence of irreversible adoption choices, a certain independence of adoption choices from price system takes place, as underlined also in the theoretical literature (see Gaffard 1990): contrary to what is suggested by the neoclassical approach, a change in the relative process may not guarantee the adoption mechanism to take place. A change in the relative price of any kind of energy source should influence the choice in favour of the more convenient source. Thus, if a carbon tax is introduced, the mix of energy inputs in production processes would generate an increase in the less costly (in terms of environmental impact) energy input, and a decrease of the relatively more costly input. In a neoclassical logic, this means that the change in the relative price of energy (e.g. an increase in carbon tax) shifts the equilibrium point towards a different mix of the two energy input factors, in favour of the more environmentally-benign energy source. Under this perspective, the introduction of renewable energy systems in the urban setting could be put in place through simple economic tools, such as taxes on adverse technologies and incentives to the use of modern and efficient technologies.

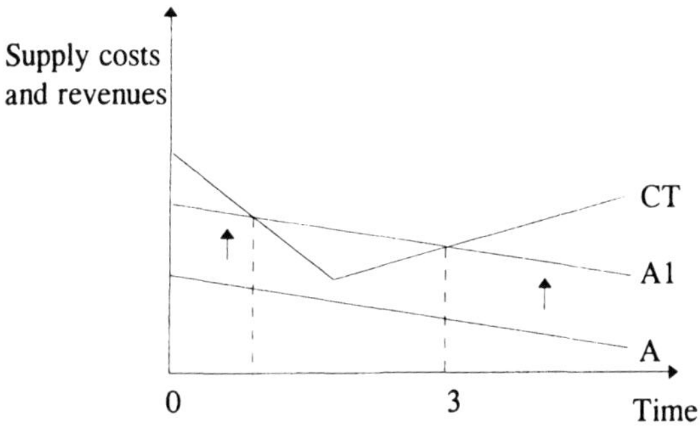

Legend:

CT Total cost of internal producers

A Advantages of the new technology

A1 Advantages of the new technology after the introduction of the carbon tax

Figure 5.3. The effects of the "carbon tax" on the equilibrium solution

In our approach, the introduction of a carbon tax increases the relative advantage of the new technology with respect to the old one: the advantage curve shifts upward to A1 in Figure 5.3, and the new technology may become profitable. However, the profitability is guaranteed for a certain time span. If the choice to innovate is for any reason postponed, even the price system which generates a higher comparative advantage is not sufficient to guarantee the innovation to take place. The neoclassical policy of changing the relative price systems does not prevent the adoption process from being heavily influenced by irreversible choices.

5.4 The Optimal Time of Adoption: the Role of Indivisibilities

As we have seen above, the adoption of renewable technologies is heavily influenced by irreversible processes, which may hamper their introduction.

However, even if the adoption takes place, there are some intrinsic mechanisms which may hamper its widespread development among potential users. Renewable energy technologies are in fact limited in number, and characterized by indivisibilities in their investments in order to build an efficient production capacity. The production of energy through different renewable energy plants is then characterized by discontinuity in the long run average cost curve which, instead of presenting the traditional U shaped curve, shows on the contrary discrete alternative solutions (Figure 5.4). Each alternative technology is characterized by a minimum average production cost. The transition from one minimum average cost point to another implies indivisible investments to achieve an efficient production capacity. This situation leads to the creation of "*local equilibria*", like point a in Figure 5.4a, which do not represent the optimal solution and the minimum energy production cost.

Figure 5.4b presents the same situation in a dynamic perspective. In a situation where the relative prices of energy sources give rise to a benefit curve like Bo, point Ao represents the maximum advantage obtained through technology on a very limited market share, or a niche where the technology is introduced and shows a higher profitability than widespread adoption, represented by Co. If we insert in the benefit curve also the advantages of a subsidy given to the new technology (or a tax on the old technology), the benefit curve of the new technology system shifts upward to curve B1. This curve shows that the widespread diffusion of the new system guarantees a higher profitability in point C1. This point, however, would never be reached if technological research processes developed by firms were not so strategic and long term oriented to foresee the greater profitability in widespread adoption, and the production could only be limited to the local equilibrium point A1 (Camagni 1996; Erdmann 1993).

a) a static view

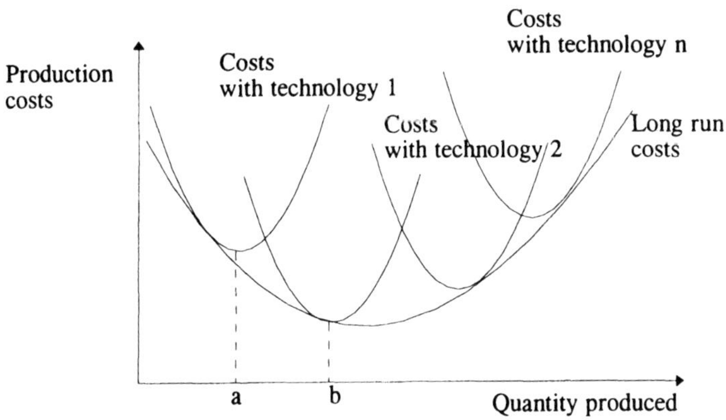

a) a dynamic view

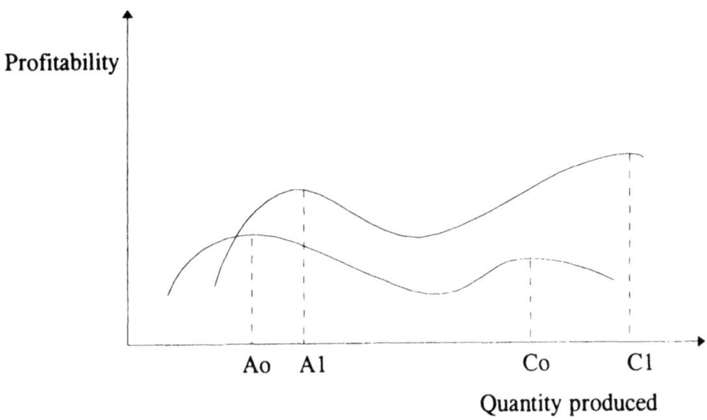

Figure 5.4. Local equilibrium solutions

5.5 The Role of Local Public Administration

As we have shown, the introduction of the new technologies is not a simple process which may be influenced by the use of traditional policies such as a change in the relative price system of energy sources.

Other policies have to be thought of, which decrease the adoption costs of the less polluting new energy technologies. One important policy, which a local public administration could put in place to help the adoption process to develop at the urban level, is the development of pilot projects. These projects would generate important advantages, by increasing context learning, creating a know-how among users, developing public awareness about the benefits of the new technology and some initial development of complementary assets around it.

In more abstract terms, pilot projects would start the new learning processes on new technologies, and especially would break the monopolistic effects of cumulative and irreversible learning processes around the old one. Moreover, this policy would start the learning processes on new technology. In conceptual terms, such a policy would shift the total cost curve downward and would decrease the slope of the relative advantage curve, leading to a relevant effect: the shift over time of the irreversible mechanisms in the adoption choice, thanks to the widening of the range of time in which the adoption is profitable (Figure 5.5). This situation favours adoption, by postponing the time in which irreversibilities in the adoption processes manifest themselves.

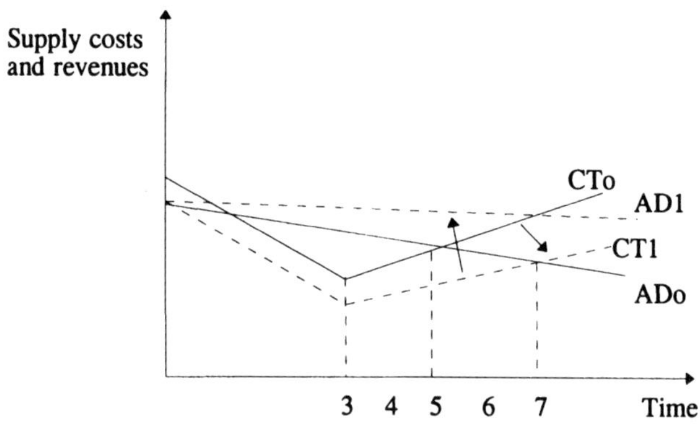

Legend:

CTo = Total costs before the implementation of the policy
CT1 = Total costs after the implementation of the policy
ADo = Total relative advantage before the implementation of the policy
AD1 = Total relative advantage after the implementation of the policy

Figure 5.5. The effects of a "pilot project" strategy on irreversible choices

Another interesting policy which could be put in place by local administrators, possibly together with the previous one, is financial support for complementary "end of the pipe" investments. The decrease of entry barriers caused by incompatibility of distribution network stardards would help the adoption process to take place. The lock-in mechanisms of the old technology would be limited, and the irreversible choice would be postponed over timc. In graph terms, this would mean a decrease in the slope of the total cost curve, and a shift from time 7 to time 8 of the no-return point in Figure 5.6a, or even its disappearance, in case the two policies would be jointly implemented (Figure 5.6b).

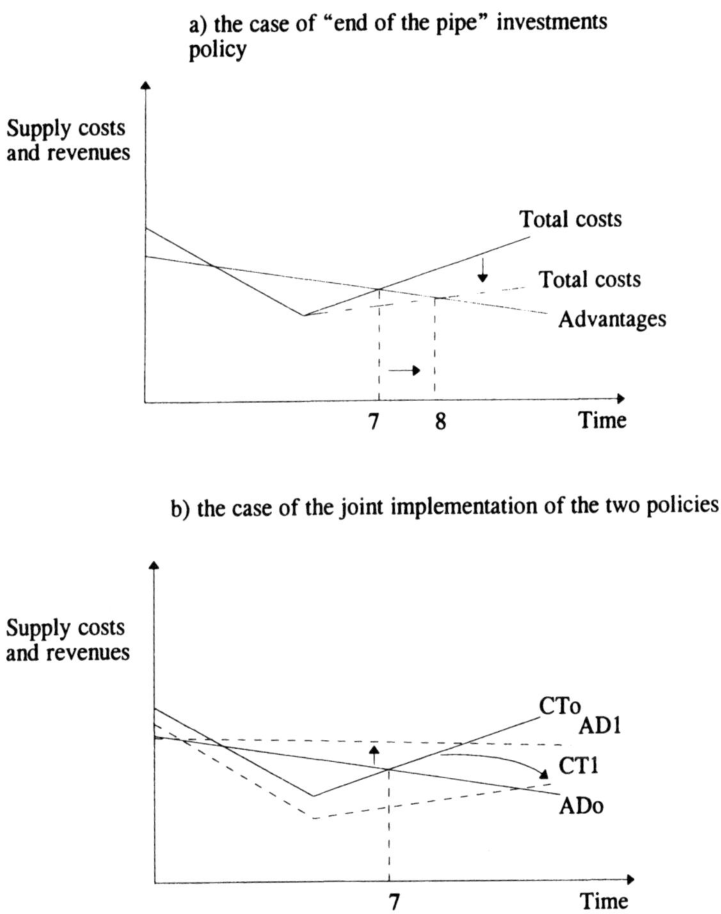

Figure 5.6. The effect of "end of the pipe investments" on the irreversible choices

5.6 The Importance of Anticipatory and Forward-looking Policies

A major and strategic conclusion stems from our analysis: the importance of anticipatory and forward-looking policies in the adoption of renewable energy technologies.

The adoption process of these technologies is fraught with many difficulties, which discourage and even tend to postpone the decision to innovate: the high indivisibility costs and risks on the new technology, difficulties to start new learning processes, and high sunk costs on the old technologies, are all elements which tend to favour a late adoption.

On the other hand, the presence of learning processes on the old existing technologies and the cumulative complementary assets around it, in the form of producers/users interface investments and distribution networks would strongly penalize cautious and postponing strategies drying out rapidly the possible advantage of the new technology. Postponing strategies may end up in impeding indefinitely the development of the new paradigm.

Far-sighted public policies should therefore adopt an anticipatory perspective, supporting early adoption and allowing the "infant paradigm" to benefit from early learning processes. As Figure 5.7 shows, if an environmentally-benign technology (EB) starts to be developed at time 0, its development may involve a limited extra cost with respect to another environmentally-adverse technology (EA). But if EB starts at time 3 while EA starts at time 0, the disadvantages in terms of private costs increase, and so does the public subsidy required to support the late introduction of the benign technology (Camagni 1996).

If this is true for industrial choices, it also holds good for spatial development trajectories in urban land use. The physical configuration of urban settlements, once established, is irreversible, unless high sunk costs are borne. For this reason, land use planning has to be carried out together with energy supply planning. Synergies and positive cross-advantages emerge from a joint strategy in the two fields, and irreversibility costs of wrong choices are avoided.

A forward-looking perspective is also convenient from the point of view of individual suppliers; in fact, an early adoption may be advantageous since it allows them to start the new cumulative learning processes before the competitors.

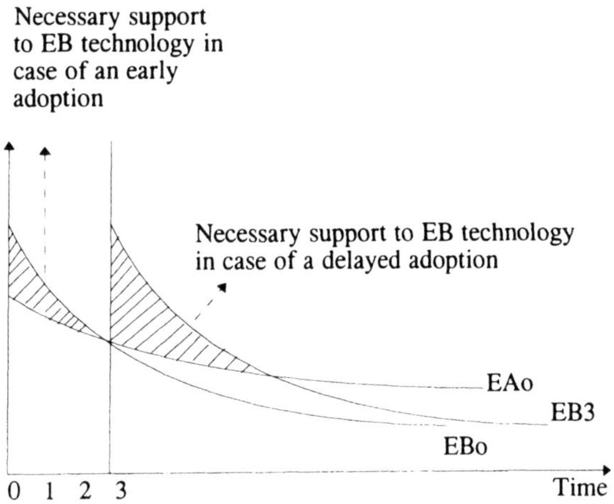

Legend:

EBo = Learning curve of the environmentally benign technology adopted at time 0

EB3 = Learning curve of the environmentally benign technology adopted at time 3

EAo = Learning curve of the environmentally adverse technology adopted at time 0

Source: Camagni (1996)

Figure 5.7. Time development of economic advantages of environmentally begin technologies

5.7 Conclusions

Although renewable energy technologies have been envisaged as an efficient and promising policy instrument to cope with environmental problems and with high energy consumption in cities, at the European level their adoption rate is still very low.

An explanation to this apparent paradox is given, when the analysis of the adoption process of renewable energy technologies is carried out in an evolutionary perspective, where time is not considered as chronological time, but is understood as the duration and continuous process of creation, as the rhythm of innovation and of cumulative learning processes, along irreversible paths. The adoption of the new energy supply technologies means a shift in the technological paradigm, a jump towards a new technological trajectory, which implies high sunk costs on previous technologies, as well as high indivisibilities on the new ones.

If this is a valid perspective, some considerations can be drawn from the analysis:

1. contrary to what seems to be the present tendency, *the adoption has to take place at an early stage*: in fact, if there is a point at which the new technology could be profitable, it is before irreversible lock-in mechanisms on old technologies, and cumulative growth of complementary assets around it become so widespread, as to reduce and even cancel its relative profitability;
2. this applies to both external and internal energy suppliers. For the former, *the optimal time of adoption, if it exists, is usually at the time when the innovation is conceived.* For the latter, the existence of sunk costs may shift the optimal time of adoption. However, also in this case, a no-return point exists, after which the new technology becomes non-profitable;
3. contrary to the standard evolutionary perspective, which considers irreversibility and lock-in mechanisms as impinging on the old technology as a consequence of the existence of sunk costs, in our case irreversibility regards a crystallization of the decision of non-adoption of the new technology. Under this perspective, *the cost of the irreversibility weighs on the new technological paradigm*, since it impedes its development;
4. the shift towards a new technological paradigm is characterized also by indivisibilities in the investments on the new technology. Their presence may not prevent the expected choice of the most efficient production point from taking place; in fact, the presence of a discontinuity in the long run average cost curve leads to the creation of "*local equilibria*", which do not represent the optimal solution and the minimum energy production cost;
5. this approach has some interesting normative consequences. The first is that traditional policies such as a tax which changes the relative prices of different energy sources (while waiting for the market to achieve a new equilibrium by itself) *may only enlarge the time span in which the new technologies are potentially profitable, postponing over time the no-return point*, but are unable to avoid the risk of irreversibilities. Points of apparent equilibrium may never be reached and decisions to adopt the new technology may never be made;
6. the last consideration has a second, and far more important, normative consequence: *it assigns new and important tasks to the public sector, whose*

policies are required to be far-sighted and anticipatory. Local administrations play the important role of developing policies which must intervene on the evolutionary mechanisms, removing any obstacles which may show up in the form of indivisibility, risks, absence of complementary assets, and the like, and above all, taking decisions about the appropriate time, before the decision of non-adoption becomes irreversible. Only with such policies, irreversibility mechanisms are kept under control, postponed over time, or may even disappear. Only with such support during their adoption process, can renewable energy technologies really become one of the possible instruments to control environmental pollution, energy use and resource depletion at the urban level.

In conclusion, the development of the renewable energy technology paradigm is hampered by strong economic elements, stemming from the evolutionary nature of technological development:

- strong indivisibilities that make the evolutionary development a discontinuous path;
- high sunk costs and adjustment costs on the previous technologies, creating obstacles for early adoption of the new technology;
- learning processes and cumulative development of complementary assets around the old technology, which work exactly on the opposite direction, penalizing a late adoption of the new technology and even making a decision of non-adoption irreversible.

6 Indivisibilities and Irreversibilities of Renewable Energy Adoption Processes: National Competition vs. Local Monopolistic Markets

6.1 Introduction

In the previous chapter an adoption choice model for energy suppliers has been presented under the assumption of a national monopolistic energy production market. However, as witnessed by the analysis carried out in Section 3.4, the recent institutional changes governing the energy market drive the energy industry towards a more open market structure. These changes are important for our theoretical model, since the institutional rules governing the market influence the trends of adoption costs and revenues, their values over time, and thus the optimal time of adoption.

In various countries the newly introduced rules leave open the possibility for local market monopolies in urban areas. Since the aim of our study is to analyse adoption trends of renewable energy technologies in urban areas, our model takes into account the particular situation of a national competition versus a local monopoly structure. Thus, we abandon the idea that the energy supplier is the only possible adopter: it is no longer true that if he does not enter the renewable energy market at the national level, nobody else will take his place.

Our main interest is to see whether, under different institutional rules at the national level, the local market has an advantage in terms of more stimuli towards innovation at the local level. Moreover, we wish to determine whether the main results of the previous monopolistic model remains the same i.e. that - in the presence of clear indivisibilities, new learning processes and specific know-how required for the new paradigm - the choice of postponing the adoption of a new technology becomes an irreversible choice.

The present chapter develops the model of the optimal adoption time in the case of a local monopolistic supplier who can decide to enter the local market whenever he decides, since nobody else will enter *the local market* in his place. However, when he enters the market, he is free to supply energy to his own local market and also to the national market, where other suppliers have already decided to enter the renewable energy technology market. The monopolistic position holds good for the local market, but not at national level, where the market is open to any interested supplier.

In this case, national competition influences the choice of the local monopolistic agent; his decision has to be based on the trade-off between the exploitation of his local monopolistic position (which allows him to postpone, or even avoid, the costs associated with the change towards the new technological paradigm) and the risk of exclusion from the national market.

The general aim of the paper is the analysis of the optimal time of adoption for a local monopolistic agent, under different hypotheses on the behaviour of his national competitors. If the result of the analysis is still that the choice of a local agent to wait to enter the market may become irreversible, this has two important consequences:

- *for the local supplier* who is excluded from supplying on the national market;
- *for the local community*, which cannot take advantage of environmental quality improvements stemming from the use of the new technologies.

In particular, the aim of the chapter is threefold:

- to analyse adoption costs and revenues over time of the local supplier under different hypotheses on the behaviour of the national agents entering the market;
- to define the optimal time of adoption under the different hypotheses;
- to suggest industrial policies in the field of energy supply at the local level.

The paper is structured as follows. In Section 6.2 we present the different hypotheses that can characterize the market at national level; Sections 6.3 and 6.4 present the optimal time of adoption at the local level under the different behavioural hypotheses on the national market, while Section 6.5 provides some policy lessons.

6.2 Different Behavioural Hypotheses on the National Market

The market structure we are investigating is presented in Fig. 6.1. The national market is open to competition in the supply of energy through renewable technologies. Energy is sold to a national carrier which distributes the energy to final users. Parallel to this market, a vertically integrated market exists at the local level, characterized by a monopolistic structure. The product supplied is homogeneous in the two markets. The relationship between the two markets is that once the local monopolist enters the renewable energy market, he can start selling energy either to the national carrier or to the final users, at a normal market price.

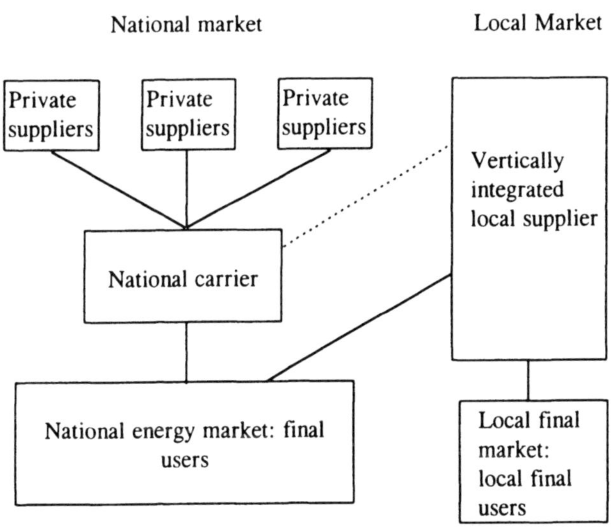

Figure 6.1. Structure of national and local markets and their interrelationship

The decision to enter the market is made by the local supplier on the basis of adoption costs and revenues, following the logic of our previous model presented in Chapter 5. The definition of the optimal time of adoption for a local monopolist depends on the institutional rules governing the national market. In our analysis three hypotheses are made on the behaviour of the agents who enter the national market:

- they can decide to cooperate with other entering agents;
- they can decide to compete, when the new technologies are not compatible;
- they can decide to compete, when the new technologies are endogenously compatible.

A common hypothesis of the three cases is that each supplier entering the market supplies energy with a particular technology, that is different from the technology used by another agent. The three cases represent three different levels of competition: the first case assumes no competition, the second shows both economic and technological competition, the third works under the assumption of only economic competition on the national market.

In the first case, the hypothesis is that cooperation takes place among the new agents entering the national market. Cooperation is in this case expected at two levels:

- at an economic level: in the energy market, an agreement may take place on the prices at which energy is supplied, and this price level remains constant over time;
- at a technological level, an agreement may be reached on the technological standards characterizing the new technologies and their complementary assets, so that they are compatible by definition.

The other two cases are characterized by competition on the energy market production: energy prices are the result of market forces, and by definition decrease over time.

The difference between the second and third case, is on the contrary, related to the technological aspects: in the second case (case b), the different renewable technologies introduced work with different standards. Standards are defined by David and Greenstein (1990) as the set of technical specifications adhered by a producer, in terms of both the core technology system, and the whole complementary assets around it. Different standards imply different complementary assets, different know-how and different learning processes, and for this reason a new competition for the definition of the new standards, which start a new technological trajectory around a specific cumulative technological path, takes place. The development of a particular standard gives rise to cumulative and self-reinforcing learning processes and specific routines so that even when weighed against a technically superior new standard, it may very well be that a first introduced Pareto-inferior standard is the one which prevails on the market (David 1985).

The second case takes into consideration the fact that different renewable technologies are non-compatible; the same standard does not apply to each technology. In the case of renewable technologies, the meaning given to the word "standard" is a general meaning, and thus the definition of non-compatible technologies may be interpreted in different ways:

- technologies which require different primary energy sources: for example, photovoltaic systems are based on different technological systems than wind energy systems. In this sense, a standard refers to energy plants based on the same technologies;
- technologies which require ad hoc complementary assets, like the district heating system which is based on a particular heat distribution grid, with respect to energy networks. In this sense, the standard means energy plants exploiting the same distribution grid;
- technologies which require additional ad hoc complementary assets, like expansion of the transmission and distribution networks. Photovoltaics or wind plants which are located near a transmission network do not require additional complementary assets with respect to the previous technological paradigm. On the contrary, photovoltaics or wind plants which are far from an already existing transmission and distribution network, or district heating systems which require in any case a new network for distributing heating, are technologies which are not immediately compatible with the old energy system. In this sense, a standard concerns energy plants exploiting the existing network grids, with no necessary physical expansion.

The model we present here aims at defining the optimal time of adoption for a local monopolist under the assumptions that the three different institutional situations presented above characterize the national market. To achieve this goal, the model analyses the discounted values of the adoption costs and advantages over time for the local monopolist, under the different assumptions made on the institutional rules governing the national market.

6.3 Optimal Time of Adoption for a Local Monopolist under the Assumption of Cooperation on the National Market

6.3.1 The reasons for the time trend of adoption costs and revenues

The first step towards the analysis of the optimal adoption choices is the definition of the time trend of the different categories of costs and revenues. Also in this case the model is built for the analysis of the adoption choice for an energy supplier, while final users are not taken into consideration. If this is the case, adoption advantages are conceived of as adoption revenues for energy suppliers.

In our perspective, the comparison between the trend over time of the discounted value of the adoption costs and revenues highlights the optimal time of adoption for a supplier. Table 6.1 contains the different reasons for the trend over time of adoption costs and revenues. Cost are divided between:

- entry barriers, which reflect the costs for each agent interested in entering the renewable energy market, and;
- exit barriers, which relate to the adoption costs for an agent who is already present in the local energy market and has to decide when the optimal time is to jump in the new technological paradigm.

Clearly, if the entry and exit costs are negligible, we are faced with a contestable market. Also in this case, the analysis is carried out for both the internal suppliers, which already act in the local market (municipal suppliers), and the external suppliers, which may enter the local market on the basis of licences obtained by the local suppliers.

The logic which explains the time trend of adoption costs and revenues is that while the local supplier waits for the optimal time of adoption, competitors enter the national market in the meantime, and cooperate (voluntarily or via legal imposition, as in the case of Italy with the law 6/92 which fixes energy prices for different categories of users) at both the economic and technological level.

Entry barriers are subdivided into indivisibilities and complementary assets. The first category of costs, *indivisibilities*, are of a different nature, and have different time trends (Table 6.1).

Table 6.1. Adoption costs and revenues over time for a local supplier in case of *cooperation* on the national market

Category of costs and revenues	Time trend	Reasons for the time trend
ADOPTION COSTS		
Entry barriers:		
a) *Indivisibilities*		
- technological	constant	
- market	decreasing	increasing returns to adoption
- organisational	decreasing	context learning
- competence	decreasing	context learning
- institutional	decreasing	context learning
b) *Complementary assets*		
- technological	decreasing	context learning
- organisational	decreasing	developed by first
- competence	decreasing	entrants
Exit barriers		
a) *Adjustment costs*		
- competence/organisational	increasing	lock-in mechanisms in the old organisational routines
b) *Sunk costs*	decreasing	cumulative depreciation of existing capital stocks
ADOPTION REVENUES		
a) *Net revenues (for internal suppliers)*	slightly decreasing	opportunity costs on the old technological trajectory and increasing returns to adoption on the new technological trajectory
b) Revenues (for external suppliers)	increasing	increasing returns to adoption on the new technological trajectory
c) *Exploitation of advantages of first adopters*	decreasing	cumulative number of suppliers in the national market

Technological indivisibilities relate to the indivisible costs that new entrants have to bear in order to build a certain productive capacity, and cannot be distributed over time. These kinds of costs remain constant over time, since at any moment in time the agent decides to join the market, he has to face the indivisible costs of large plants and of large production capacity, in order to achieve a profitable production.

Organizational indivisibilities represent the costs associated with the high organizational costs that have to be borne once a new production line or plant is developed. Every moment in time the decision to adopt is made, organizational costs have to be borne; their value over time remains constant.

Market indivisibilities regard the critical mass of adopters that each technology has to achieve in order to develop a cumulative and self-sustained adoption process. These kinds of costs decrease over time. The presence of other suppliers at the national level supports the achievement of a critical mass of subscribers. The late-comer reaps advantages from the existence of an already established demand in the national market, which positively influences the decision to adopt at the local level.

Competence indivisibilities regard the level of know-how which is required for the development of the new technology. While time passes, these costs are expected to decrease because of the cumulative learning processes which are generated by R&D activities within the environment in which the potential agents act. Over time, the level of competence which the firm itself has to develop in order to innovate, decreases more and more. The firm finds the necessary competence on the market, developed by first comers, and can buy and exploit it. This mechanism reinforces itself while time passes, and thus competence indivisibilities decrease over time.

Institutional indivisibilities relate to the need of an institutional support, via lobbying activities, to develop a public and social awareness of the importance of the new technologies. These costs decrease over time; it is foreseeable in fact that the awareness among local, regional and national policy-makers about the importance of new energy technologies increases over time, driven by cumulative consciousness about the social costs associated with energy consumption.

The second group of costs which constitute entry barriers to a new technological paradigm is represented by *complementary assets*. The shift towards a new technological paradigm implies the reorganisation of the development pattern around a radical innovation. However, around the core technology which drives the change, a number of interrelated technological upgradings in all complementary assets is required. This is even more true when technological paradigms are characterized by an increasingly complex and systemic nature: in these cases, the radical innovation on which the new paradigm rests, influences the spectrum of technological complementary assets (Antonelli 1995).

The same pervasive and drastic change is required in the organizational structure of suppliers and customers: they have to break the traditional organizational routines and build new ones around the new technological processes. Needless to say that the new technological paradigm imposes a new learning path even in all complementary assets of the core technology. This group of costs decreases over time and the presence of national suppliers, who enter the market, creates a cumulative knowledge around the new core technology by generating a diffused and cumulative know-how around the new paradigm. Since the technologies are compatible and thus work around the same technological standards, late-comers gain the advantage of the already existing know-how, and the learning processes already stimulated by early entrants.

If the analysis is carried out in the perspective of an already existing supplier, other kinds of costs have to be analysed and added to the previous ones, and may be classified as exit barriers. Two categories apply to this group.

The first are the *adjustment costs* which are the counterpart of the high organizational indivisibilities that have to be borne by new entrants, when they decide to innovate. Adjustment costs regard in fact the organizational and managerial sphere of a firm, which, in the case of already existing suppliers, does not need to be structured ex-novo, but requires adjustment to the new paradigm. Sometimes however, even the reorganization of labour and functions is expensive. In the perspective of the evolutionary school, this element is one of the main obstacles to the introduction of a new paradigm. Organizational routines may lock the firm in the previous technological trajectory and limit both the willingness and interest for any other techno-organizational solution Nelson and Winter (1997) have strongly emphasized the existence of organizational routines in a firm, around which the whole management of the firm is organised. Organizational routines are based on learning processes, which are cumulative in nature, and which for this reason reinforce themselves while time passes. In the dynamic perspective, firms are locked-into the organizational routines that have been put in place, and the introduction of a new paradigm is from this perspective more and more costly, while time passes.

The last category of exit barriers are the *sunk costs* related to the old technology, such as the incomplete depreciation of previous capital stocks. The full exploitation of the previous technological plants is something a firm can try to achieve; but it also inhibits the move towards the new technology in the short run.

These costs may justify a conservative strategy, since they decrease over time. A late adoption is from this perspective a more profitable solution.

The counterpart of these costs are the *advantages* that the introduction of a new technology generates. Also in this case, as in Chapter 5, the perspective is the one which holds for suppliers rather than for final users. If this is the case, the advantages we are speaking about are adoption revenues. For internal suppliers the revenues analysed are the net revenues stemming from the difference between the absolute revenues of the new technologies and the opportunity costs to remain on the old technological trajectories; these latter increase over time. The absolute revenue of new technologies has a positive trend over time, since renewable technologies are networked technologies, whose revenue is influenced by increasing returns on adoption. In particular, as Arthur underlines (Arthur 1988), technologies may become more and more attractive, the more they are developed: "increasing returns to adoption" exist and rise from several sources (Arthur 1988):

- learning by doing (Rosenberg 1982): the more a technology is adopted, the more it is used, and the more is learnt about it;
- network externalities (Katz and Shapiro 1985, 1986, 1992): the more other users there are, the more lickely it is that the adopter benefits increase because of a greater knowledge around that technology;
- informational increasing returns (Arthur 1988); a technology that is more adopted enjoys the advantage that is better known.

The sum of these two counter-effects is difficult to predict, although we are more inclined to think that the opportunity costs of the old technology prevail on the increasing returns to adoption characterizing the revenues of the new technology. For this reason, it is plausible to expect that the revenue curve will slightly decrease.

Under the assumption of fixed prices of energy produced via renewable technologies, as in the case of cooperation at the national level, revenues for external suppliers are by definition not influenced by opportunity costs of the old technology, but only by increasing returns on adoption, and thus certainly increase over time.

Another advantage associated with the introduction of the new technology is the typical advantage of the first adopter. The condition of being the first supplier to introduce such a product innovation would provide important returns as it would help exploit general psychological elements, such as curiosity and excitement about the novelty. These kinds of advantages hold for both already existing energy suppliers and for new ones. While time passes, these advantages decrease over time, because other actors meantime enter the market.

The analysis of the optimal adoption choice is the result of the comparison between costs and revenues; from their difference the adoption profitability and its time trend are determined. As in the case of a monopolistic market, the profitability for the 'external suppliers' is an absolute profitability of the new technology, while for 'internal suppliers' the profitability on which they take decisions is relative, which discounts the opportunity costs of remaining on the old technology.

In the next section, the optimal time of adoption is envisaged in the case of cooperative agreements on the national market.

6.3.2 Irreversibilities in the case of cooperation on the national market

The analysis of the discounted values of adoption costs and revenues enables us to analyse the optimal time of adoption of renewable energy technologies. The perspective under which the optimal adoption time is analysed is twofold:

- the perspective of an 'internal supplier' i.e. a supplier who is already in the local market and looks for the optimal time to move towards the new technological trajectory (Figure 6.2a);
- the perspective of an 'external supplier', who knows that at any moment in time he decides to enter the local market, he will be the first one, and in this perspective, he looks for the optimal time of adoption (Figure 6.2b).

In Figure 6.2a net revenues decrease over time because of the opportunity costs that 'internal suppliers' have to discount from remaining on the old technological trajectory. On the contrary, costs first decrease, because of:

a) decreasing market indivisibilities;
b) decreasing competence and institutional indivisibilities thanks to context learning;
c) decreasing costs related to the context learning around the complementary assets;
d) depreciation of the already existing capital stock on the old technology. For 'internal suppliers' however, while time passes, adjustment costs increase and the adoption cost curve increases after a certain time span, as a result of lock-

in mechanisms around the old technology, which start to generate their negative effects.

The result of these trends is shown in Figure 6.2a: a time span exists during which the new technology is profitable, after which adjustment costs from the old to the new technology are so high that they overcome the relative advantage. The profitability of the new technology varies according to the fact that: the old capital stock is entirely depreciated; and the negative effects of adjustment costs prevail on the positive effects of the increasing returns on adoption and on context learning.

The main result in this case is similar to the case of a monopolistic market at national level: if the new technology is to be profitable, internal local suppliers have to enter the market before the organizational and managerial routines lock-in the potential adopter in the old technological paradigm. After a certain time span, the technology becomes no longer profitable, and the choice of non-adoption becomes irreversible.

This situation is similar to that of a monopolistic market for 'internal suppliers', and also in this case optimal adoption time, depends on the slope of the revenue costs in comparison with the slope of adoption costs. Also in this case, there are three possibilities:

- the time derivative of costs exceeds that of revenue. In this case, for already existing energy producers, the optimal time of adoption is when the old capital stock has been completely depreciated (Figure 6.2.a). If an optimal time of adoption were to exist, this would be at time 2, when the difference between adoption costs and revenues is wider. The early adoption does not seem to be the right strategy for already existing energy suppliers: the sunk costs associated with their previous technology oblige them to wait until the depreciation process has ended;
- the time derivative of costs is equal to that of adoption revenues. In this case (Figure 6.2.b), the optimal time of adoption is in the range between time 0 and 2. Within this time span, the choice is optimal, though the technology is still profitable until time 3;
- the time derivative of costs is lower than that of revenues. In this case (Figure 6.2.c), the optimal point of adoption is time 0.

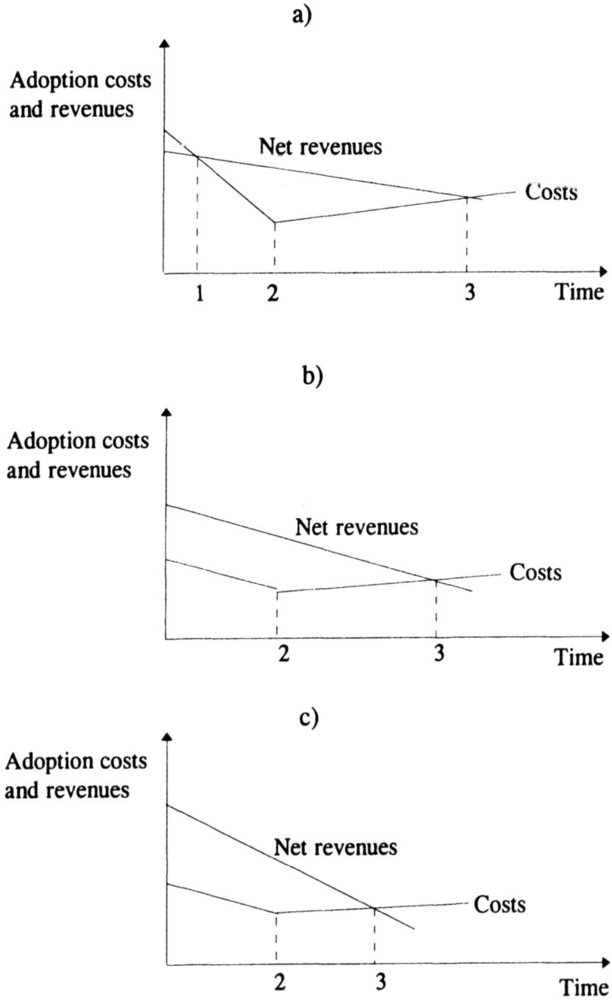

Figure 6.2. Local adoption costs and revenues for a local monopolistic agent under the assumption of a cooperative national market: the view of an internal supplier

In all cases however, a forward-looking policy is necessary. The local monopolistic agent has to foresee that an early entrance in the national market allows him to achieve greater profitability from the adoption, despite the opportunity costs associated with a choice of remaining on the old technological trajectory, in a protected market.

Another situation characterizes the decision to enter the market for an 'external supplier'. Figure 6.3 presents such a case: the revenue curve increases over time, since external suppliers benefit from the increasing returns to adoption, while costs in this case constantly decrease, because of increasing context learning and increasing returns on adoption. A late adopter takes advantage of the decreased costs of the necessary learning processes and cumulative know-how, which are borne by first entrants at national level. In this perspective, the optimal adoption strategy is to wait, and leave the others to decide to bear the costs of the new learning processes related to the new technological paradigm.

This also means that cooperation generates positive factors, which increase over time: late adopters exploit lower adoption costs associated with learning mechanisms already put in place by first comers. In this strategy, the risk is that the decision to enter the market at the local level is permanently postponed, because of the greatest profitability level reached through a "wait and see" strategy.

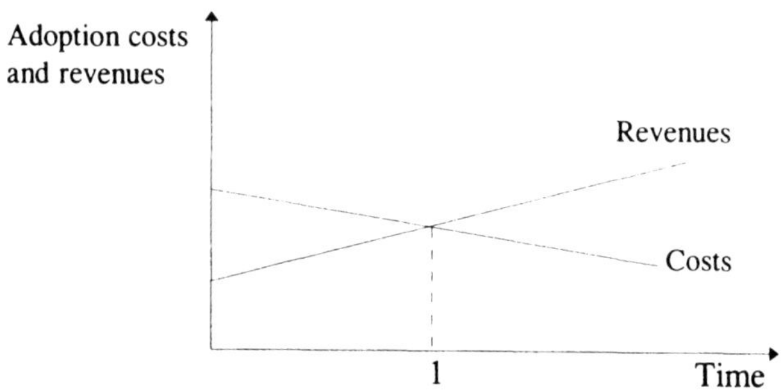

Figure 6.3. Local adoption costs and revenues for a local monopolistic market under the assumption of a cooperative national market: the view of the external suppliers

6.4 Optimal Time of Adoption for a Local Monopolist under the Assumption of Competition with Non-Compatible Technologies on the National Market

6.4.1 The reasons for the time trend of adoption costs and revenues

The same analysis is carried out by changing the hypotheses on the institutional rules governing the national market. In this case, the hypotheses made are:

- in the national market, agents supply energy under competitive rules. No price agreements exist in this case, as will be the case of energy sold to large users after the introduction of the European Community Directive;
- the technologies work under different technological standards. By definition, therefore, they compete, since the standard chosen for the first introduced technology will create a learning process around it, and will instead increase the costs of complementary assets around the other possibly adopted technologies. In the renewable energy technology market, this is mostly the case.

Table 6.2 contains the same breakdown of adoption costs and revenues and their time trends: these will change, because of the different assumptions made about the rules governing the national market. They are analysed under the assumption that the local monopolistic agent remains free to choose a technology which is different from the ones introduced at national level, whenever he decides to enter. The changes in the cost trends over time with respect to the previous case relate to:

- *market indivisibilities* The new technologies have different and non-compatible standards; this assumption means that the different technologies are developed by different suppliers, which do not interact because of technical incompatibility. The first-comers cannot in this respect generate a critical mass of users for other technologies, which are technologically incompatible. For this reason, the market indivisibility remains constant over time: at any moment in time the local supplier enters the market, he has to attract a critical mass of users by himself;
- *complementary assets* Because of the technical non-compatibility of the different new technologies, the development of complementary assets around

the first technology introduced by first-comers does not help the development of learning processes around another technology, but even lock-into the system in the first choice made, because learning processes and cumulative competences are developed around the first standard introduced (Arthur, 1988). When time passes, the costs associated with the complementary assets of another new non-compatible technology, increase. In this respect, the technical Pareto-superiority of the late technology does not guarantee its development: as the literature on standard competition underlines, learning processes and sunk costs on the previous technology introduced can be so high that the technological advantages of the late technology may be insufficient to start the new paradigm (David 1985; Arthur 1990; David and Greenstein 1990);

Changes take place also in the time trend of the revenues with respect to the case presented in Section 6.3. In particular, while the reasoning on the adoption revenues changes, since:

- the *net revenue for internal suppliers* decreases because of two effects: a) opportunity costs of the old technology, which increase while time passes; b) energy prices which decrease on the national market. Moreover, internal suppliers may put in place price cross-subsidy mechanisms between the local and national market, in favour of the latter: lower prices may be fixed at the national level with respect to competitors, because they are subsidized by high prices at the local level; since national energy prices tend to decrease while time passes, the subsidy required at the local level has to increase, if the revenues of the local supplier have to remain constant. Since the local monopolistic prices cannot increase for ever, the revenue of the local monopolist decreases in time also for this reason;
- the *revenue for external suppliers* decreases because of the price competition effect on the national market;
- since the new technologies are non-compatible, the *advantages of the first adopter* remain constant over time. At any point in time the 'late-comer' decides to enter the market, he will exploit curiosity and excitement about the novelty.

As already shown in previous cases, the comparison between adoption revenues and costs leads us to the definition of the profitability level, and of the possible time of adoption.

Table 6.2. Adoption costs and revenues over time for a local supplier in case of *competition among non-compatible technologies* on the national market

Category of costs and revenues	Time trend	Reasons for the time trend
ADOPTION COSTS		
Entry barriers:		
a) *Indivisibilities*		
- technological	constant	
- market	constant	
- organisational	constant	
- competence	decreasing	context learning
- institutional	decreasing	context learning
b) *Complementary assets*		
- technological	increasing	learning processes on the first technology introduced
- organisational	increasing	
- competence	increasing	
Exit barriers		
a) *Adjustment costs*		
- competence/organisational	increasing	lock-in mechanisms in the old organisational routines
b) *Sunk costs*	decreasing	cumulative depreciation of existing capital stocks
ADOPTION REVENUES		
a) *Net revenues (for internal suppliers)*	decreasing	opportunity costs on the old technological trajectory decreasing energy prices on the national market
b) Revenues (for external suppliers)	decreasing	decreasing energy prices on the national market
c) *Exploitation of advantages of first adopters*	constant	

6.4.2 Irreversibilities in the case of competition on the national market with non-compatible technologies

Figure 6.4 presents the synthetic results of the time trends of costs and revenues for the "internal supplier" (Figure 6.4.a), and for an external supplier (Figure 6.4.b).

a) internal suppliers

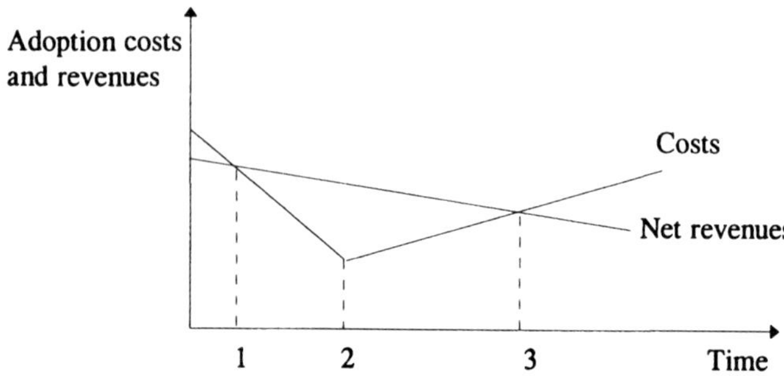

b) internal suppliers

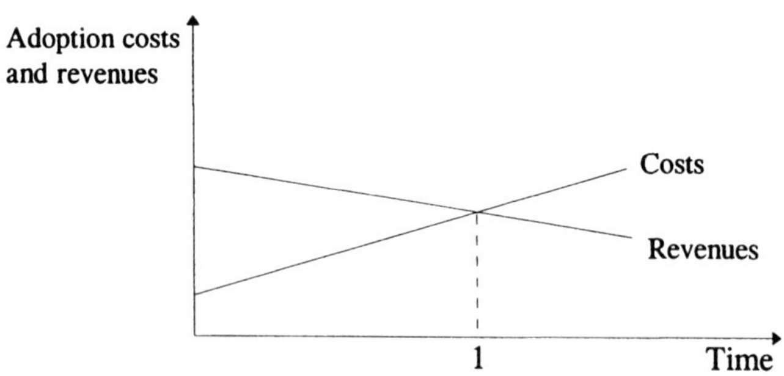

Figure 6.4. Local adoption costs and revenues for a local monopolistic supplier under the assumption of a competitive market among non-compatible technologies

The internal supplier's view

With regard to the net revenue curve, this shows a negative slope, because of the opportunity costs associated with the permanence of the supplier in the old technological trajectory, and because of the decreasing energy prices on the national market.

The cost curve is the result of different trends over time of the different categories of costs: a decreasing component, due the presence of context learning and of sunk costs on previous technology, a constant component of indivisible costs which are independent of the adoption moment, and an increasing component on complementary assets, due to increasing mechanisms of locking-into the previous organizational structure.

Figure 6.4 summarizes these trends. For 'internal suppliers', the optimal time of adoption depends on the slope of the adoption costs in comparison with the slope of the adoption revenues. Also in this case, there are three possibilities. In all three cases, after a certain moment in time, the costs of change become so high, that the choice not to enter the market becomes irreversible because the technology is no longer profitable for the supplier, who is excluded from the market.

It is reasonable to expect that sunk costs on the incomplete depreciation of already existing capital stock influence the total costs over time more than the mechanisms of continuous learning on old technology. If this is the case, the total cost curve assumes a negative slope until the previous capital stock is completely depreciated (at time 2 of Figure 6.4a). At that time, the adoption costs of the new technology start to increase, influenced by the learning processes around the old technological trajectory. The negative slope of the cost curve is in any case not very deep, since it is the result of the sum of an increasing component and a decreasing one.

The external supplier's view

Figure 6.4b represents the case of an external supplier. While the revenue curve has a decreasing trend, due to the decreasing energy prices on the national market, the costs curve increases in time, due to lock-in mechanisms in the first technology chosen by the first entrant. Technological competition is at the basis of these mechanisms since it influences the complementary assets and the context learning of the other new technologies. The result is an adoption cost curve which constantly increases over time (Figure 6.4b).

The results are quite interesting also in this case. The optimal adoption time for a local potential supplier is time 0, when the technology is introduced. If he waits, adoption costs will quickly become greater than the revenues, and the technology will become no longer profitable after time 1. The optimal adoption time then is the moment at which the technology may be put on the market. A forward-looking strategy is necessary also in this case: if a potential supplier can enter a local market as a monopolist, the optimal choice is to enter very quickly. In this way, he can exploit the high revenues of the new technology in the national market, which inevitably decrease as time passes. Moreover, he can avoid increasing adoption costs linked to a postponed entrance.

6.5 Optimal Time of Adoption for a Local Monopolist under the Assumption of Competition with Endogenously Compatible Technologies on the National Market

6.5.1 Reasons for the time trend of adoption costs and revenues

The last case considered is the one characterized by a national competitive market of compatible technologies. Under this assumption, the different costs and revenues change their trend over time (Table 6.5).

With respect to competition with non-compatible technologies, the cost components which change their time trends are those related to complementary assets. Since new technologies are endogenously compatible, the development of those technologies by other national market agents, decreases the costs associated with the know-how and the learning processes of complementary assets for late adopters.

The difference to the previous case of competition among non-compatible technologies holds good also for one specific component of the revenue curve. The component which changes is the advantage to be the first adopters, which decreases over time. Since the technologies are compatible, the first adopters' advantages are exploited by the first innovator: while time passes, and other compatible technologies are introduced, these advantages inevitably decrease.

The other two components, i.e. the revenues (net and total) for internal and external suppliers, have the same decreasing trend over time, although less emphasized. In fact, the decreasing trend is in this case alleviated by the existence of increasing returns to adoption of the new technology, stemming from the assumption that these technologies are compatible technologies.

Table 6.3. Adoption costs and revenues over time for a local supplier with *competition among compatible technologies* on the national market

Category of costs and revenues	**Time trend**	**Reasons for the time trend**
ADOPTION COSTS		
Entry barriers		
a) *Indivisibilities*		
- technological	constant	
- market	decreasing	increasing returns to adoption
- organisational	constant	
- competence	decreasing	context learning
- institutional	decreasing	context learning
b) *Complementary assets*		
- technological	decreasing	context learning
- organisational	decreasing	
- competence	decreasing	
Exit barriers		
a) *Adjustment costs*		
- competence/organisational	increasing	lock-in mechanisms in the old organisational routines
b) *Sunk costs*	decreasing	cumulative depreciation of existing capital stocks
ADOPTION REVENUES		
a) *Net revenues (for internal suppliers)*	decreasing	opportunity costs on the old technological trajectory and increasing returns to adoption on the new technological trajectory decreasing prices on the national energy market
b) Revenues (for external suppliers)	slightly decreasing	increasing returns to adoption on the new technological trajectory decreasing prices on the national energy market
c) *Exploitation of advantages of first adopters*	decreasing	cumulative number of suppliers

6.5.2 Irreversibilities in the case of competition in the national market with endogenously compatible technologies

In case of competition on the national market among endogenously compatible technologies, the time trend of costs and revenues adoption is the one represented in Figure 6.5.

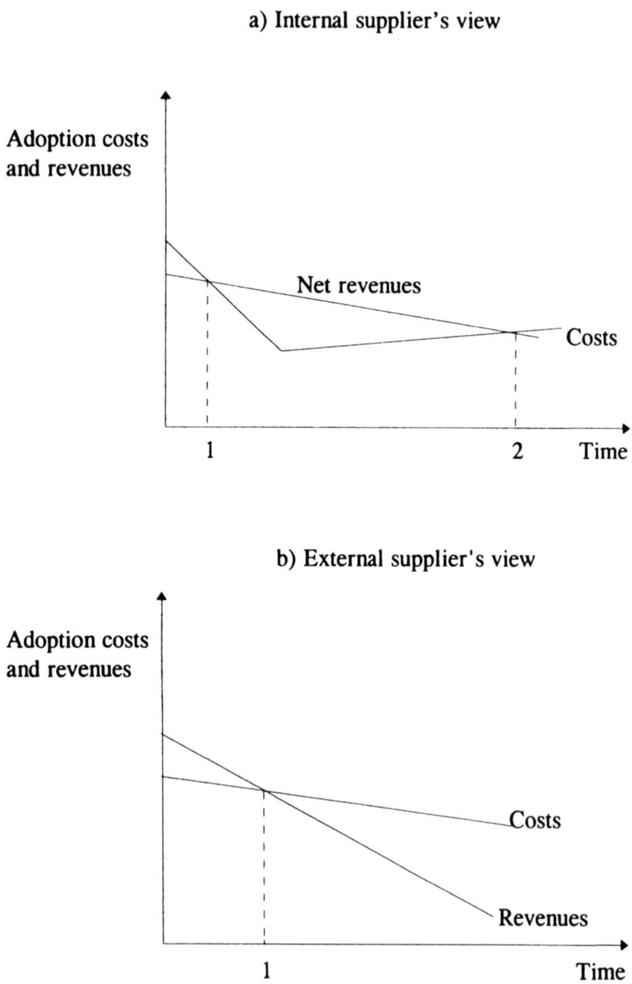

Figure 6.5. Local adoption costs and revenues in a competitive market of endogenously compatible technologies

The internal supplier's view

Figure 6.5.a shows the case of costs and revenues for an internal supplier. The situation presented shows again that the optimal time of adoption, if it exists, depends on the slope assumed by adoption costs and revenues. In any case, the choice of adoption has to be made within a certain period of time: after this period, the technology is no longer profitable, and the choice not to enter the market becomes irreversible.

The external suppliers' view

A different result is achieved if the perspective is the one of the external supplier. The revenue curve is decreasing, although less than in the case of competition at both the economic and the technological level. Costs decrease too, and the solution depends on the slopes of the two curves. However, it is plausible to assume that the negative effects of a competitive market structure, which negatively influence the slope of the revenue curve, are more effective than the positive decreasing effects on costs. If this is true, the result is the one presented in Fig. 6.5.b, and again the need for a short term choice of adoption emerges.

6.6 The Influence of National Policies on the Optimal Adoption Time at the Local Level

The main result of our theoretical exercise is the emergence of irreversible choices of non-adoption which may lock local suppliers in the old technological trajectories. As envisaged in the introductory section, the consequences of such irreversible mechanisms are of a private and social nature. The private nature is related to the loss of the local agent to enter the national market, and thus to limit his business to the local monopolistic area. The social nature of such a choice is that a loss in terms of social welfare is determined for the local community, which does not benefit from the advantages stemming from the development of less polluting technologies.

The only theoretical case in which irreversible mechanisms do not take place is where an external supplier enters the energy market, while a cooperative structure charactersizes the national market. In this case, the positive factors stemming from the costs assumed by first innovators, justify a strategy of late entrance, also for the late comers.

National authorities own important instruments to govern the market, even at the local level. These instruments may be divided into two main groups:

- price policies, aiming at regulating the relative profitability of new technologies with respect to more conventional energy systems;
- technological policies, aimed at keeping adoption costs of new technologies, and especially their increase over time, under control.

In the next sections, we present the effectiveness of these policies on the irreversible mechanisms. The main result is that, in case of high degrees of competition, single policies are not sufficient to avoid the irreversible mechanisms from taking place. In the case of both economic and technological competition, single policies only postpone the moment in which the new technology becomes non-profitable. As in the case of a national monopolistic market, even a relative price policy in favour of environmentally benign technologies does not solve the problem of opportunity costs for remaining on the old technological trajectory. Combined policies should be put in place in order to keep under control:

- the high sunk and adjustment costs on conventional technology, which push towards a late adoption;
- the learning processes and cumulative development of complementary assets and the decreasing revenues of a competitive market which would require an early adoption.

6.6.1 National price policies

A first group of policies are the relative prices at which energy has to be sold to the national monopolistic carrier (or to the final user), which influences heavily the revenues of the new technologies, and thus their profitability.

A relative price policy in favour of the introduction of the new technologies would shift the relative revenue curve upward, thereby postponing the moment in which the technology is no longer profitable for the local supplier.

Figure 6.6a shows the case of a cooperative market, or a case of competitive market of compatible technologies. In this case, the relative price policy is able to maintain a certain level for profitability of the technology. Instead, in the case of a competitive market structure where competition takes place at both an economic and technological level, such a policy would only postpone over time the moment in which the local monopolist would be excluded from the new market, as Figure 6.6b shows.

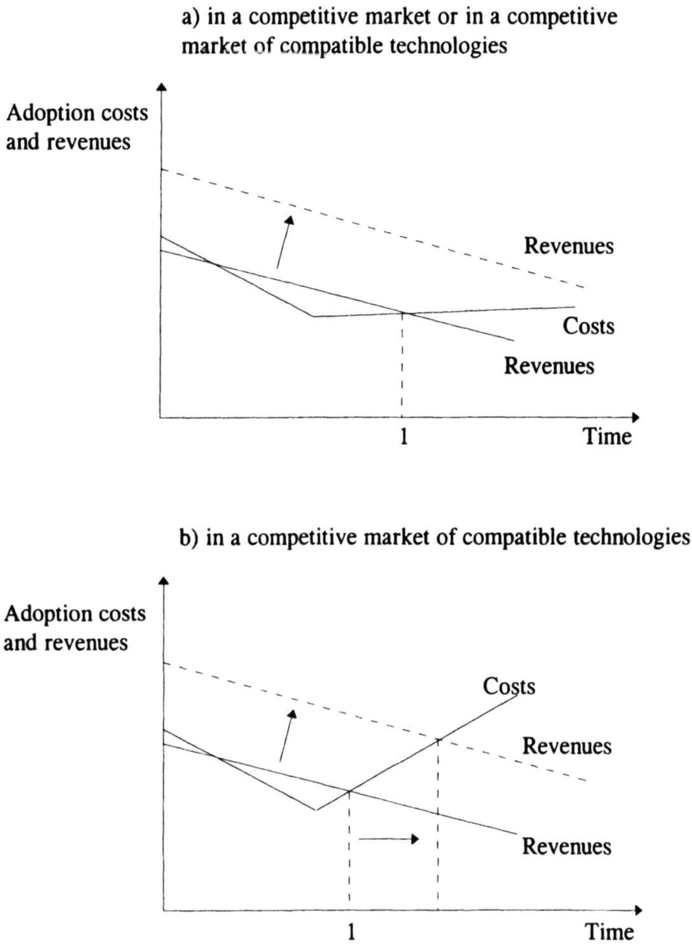

Figure 6.6. Effects of a relative price policy on the adoption moment for an internal supplier

The effect of these policies on an external supplier depends on the institutional rules governing the national market:

- in a cooperative world, this policy would anticipate the moment when the technology becomes profitable (Figure 6.7a), since it would decrease the advantage of having adoption costs paid by others. The positive factors provided by other agents entering the market at national level and thus paying the costs of adoption, are compared with an increasing profit of the technology which attracts the potential agents to enter in a more limited time span;
- in a competitive world, these policies postpone over time the moment of adoption, as in the case of internal suppliers. However, in both cases the increase in the relative profitability of the new technology does not cancel the moment at which the non-adoption choice made by a local supplier becomes irreversible (Figure 6.7b and 6.7c).

Another similar economic instrument in the hands of national authorities is the level of subsidies in favour of this technology. Putting these kinds of policies into place would have exactly the same positive effect as on the price policies shown in Figure 6.6 and 6.7.

6.6.2 National technological policies

Other kinds of policies which are extremely important in case of a change in the technological paradigm, are related to the choice of the technological standards for the new technologies, when competition takes place among non-compatible technologies. These policies become important tools in influencing the optimal adoption time. If the interest is to help the adoption of new technologies, the choice of a common standard would decrease the adoption costs in terms of complementary assets. As Figure 6.8a shows, this would decrease the costs curve and would shift the moment in which the technology is no longer profitable over time.

This is also true for 'internal suppliers' where the point at which the technology is no longer profitable, is postponed (Figure 6.8.b).

Based on the same logic, the choice of non-compatible standards becomes an efficient tool to prevent local potential supplier from entering the market. Here, in fact, the adoption costs will increase, and will inevitably reduce the time during which the technology is profitable (Figure 6.9).

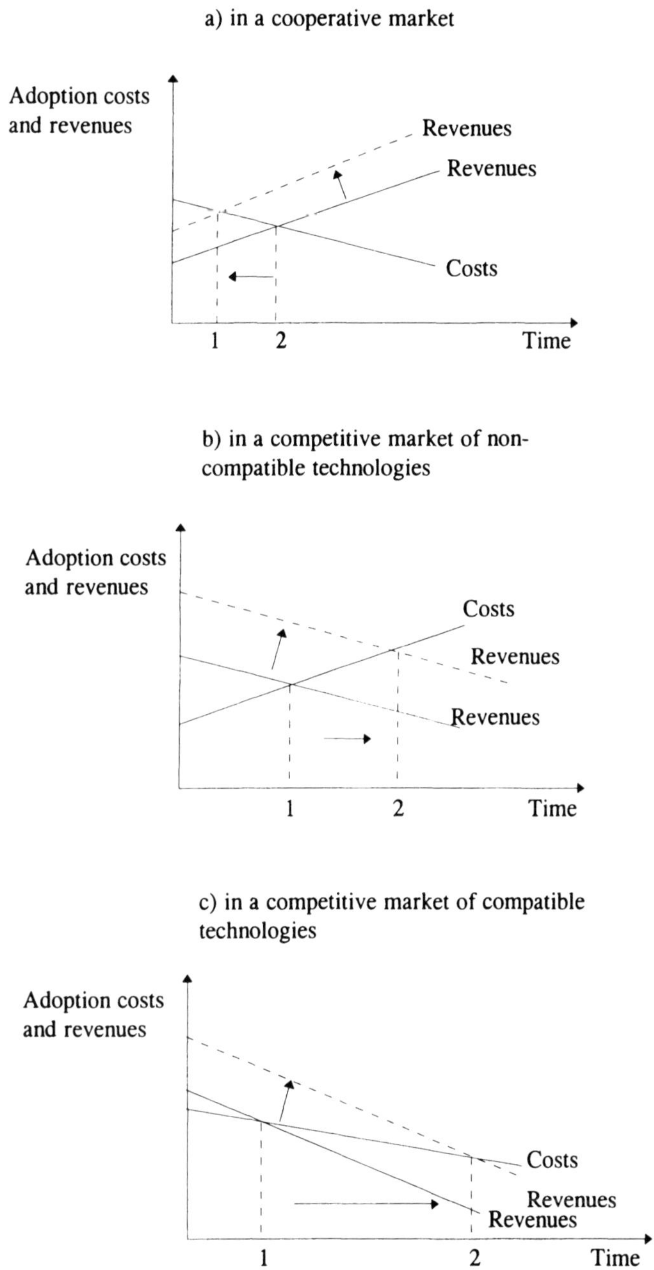

Figure 6.7. Effects of a relative price policy for an external supplier

a) Internal supplier's view

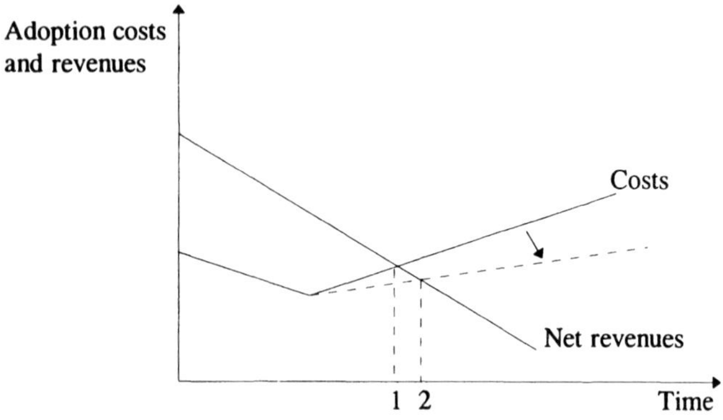

b) External supplier's view

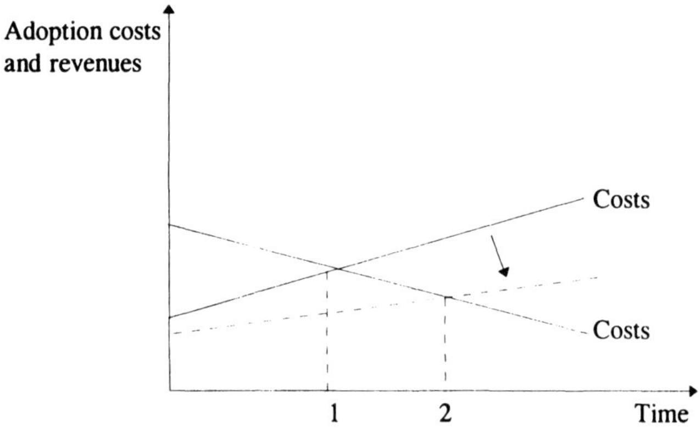

Figure 6.8. Effects of national technologies policies on the optimal adoption moment

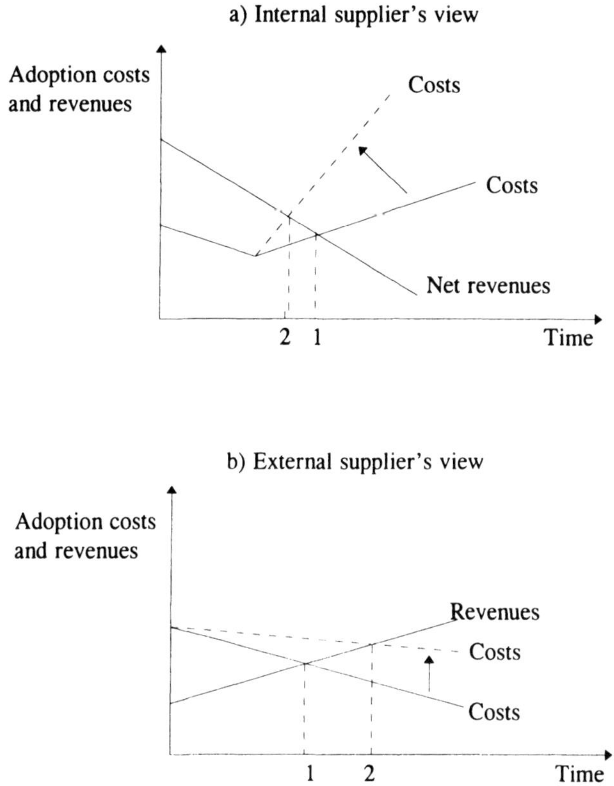

Figure 6.9. Effect of a non convergent standard approach in a cooperative market

6.6.3 Combined policies

The result of the policy tools presented above demonstrates that the most powerful tool is the use of both kinds of policies at the same time. If price and technological policies are adopted at the same time, they may provoke a more important effect, since they move at the same time both the relative revenue and cost curves. The joint effect generates even more far-reaching conclusions, since it may even permanently postpone the point at which technology is no longer profitable (Figure 6.10). Combined policies can control two aspects at the same time:

- price policies can keep the economic competition under control;

- technological policies can keep the evolutionary mechanisms under control, by removing any obstacles which may show up in the form of absence of complementary assets.

Needless to say, the same policies may become a useful instrument for keeping local suppliers out of the market, if used differently. For this reason, it may become rather important to keep the two policies in the hands of a national authority, in order to avoid a biased interference in the market.

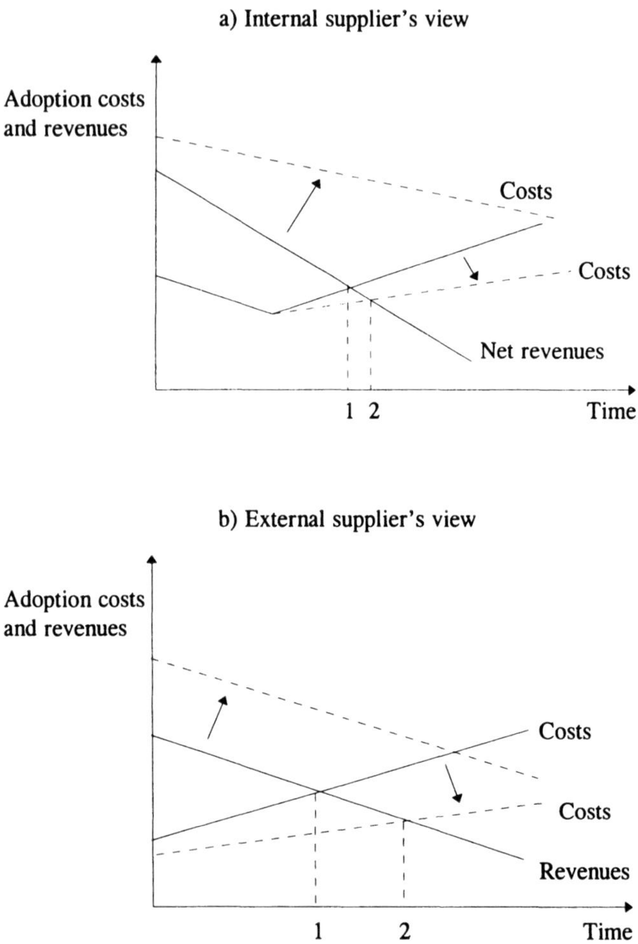

Figure 6.10. Effects of a combined policy action

6.7 Conclusions

Some interesting results are obtained from the analysis, which will be briefly summarized here:

- Not to enter the market is a very risky strategy. In the long term, in fact, entrance to the market becomes non-profitable, and thus irreversible. Even in the case of a local monopolistic position, which prevents external competition, the choice of non-adoption is not a wise one. Monopolistic advantages have to be compared with a long-term risk of being excluded from the national market, or of external forces (like institutional rules) obliging the introduction of renewable technologies. Farsighted strategies are necessary to avoid the risk of being excluded from the market.
- The importance of forward-looking policies is more valid, the higher the degree of competition. As Figures 6.6 and 6.7 show, a higher level of competition determines a shorter time span in which the decision to adopt has to be made: the non-return point of the decision to enter the market takes place in a more limited time span.
- On the contrary, only one case exists where the decision to postpone the adoption is worthwhile and that is in relationship to national cooperation. Here, from the perspective of the external supplier, the decision to wait offers greater profitability from the introduction of the new technology. In fact, if national agents cooperate, they bear the indivisible costs and the cost of the learning processes associated with the new technology. As time passes, the late entrants gain positive input into the cooperation agreements and the related cost advantages. This holds true for the external supplier's view. Internal suppliers, however, have to bear the costs associated with the learning processes of the old technology once the old paradigm is relinquished; this is quite independent of the institutional rules governing national markets. These costs cannot be transferred to any other agents and for this reason, they limit the time span in which the decision to adopt has to be taken.
- If the national market is in a competitive position, the optimal time of adoption at the local level is in most cases the moment at which the innovation is introduced in the national market. When the new technologies are non-compatible, early adoption allows the supplier to impose his standards through the development of learning processes associated with his technology. He thereby avoids increasing costs related to the growing complexity of

complementary assets over time, if the new technologies are endogenously compatible. Early adoption allows him to avoid losses stemming from high revenues, higher than the opportunity costs associated with a late adoption.

- Different policy tools at national level exist to influence the local decision to adopt. Two groups of policies have been envisaged: price policies and technological policies. Price policies can be used to increase the relative profitability of the technology. In the case of the competitive world, the advantages of a new energy technology may lead to a postponement of the moment in which the decision not to adopt becomes irreversible for the local agent. In the case of a cooperative world, the higher relative advantages accelerate the moment at which the technology becomes profitable, thereby limiting the time a local supplier has to take a wise decision.
- Technological policies are related to the technological standards of the new technologies, and are thus crucial when technologies are non-compatible. Common standards among new technologies and between new and old technologies drastically decrease the adoption costs of new technologies and postpone the moment at which the decision not to adopt becomes irreversible. Using the same logic, national authorities choices of technologically-incompatible standards may reduce the time in which the technology is profitable, and thereby anticipate the moment in which the decision to adopt is possible.
- A combined action of technological and economic policies in favour of the adoption of new technologies constitutes a rather important effect. By acting on both the adoption and cost curves, the combined policy effect even avoids the moment in which the technology is no longer profitable, thus also the moment at which the decision of non-adoption becomes irreversible for a local potential supplier.

7 Indivisibilities and Irreversibilities in Urban Renewable Energy Strategies: a Comparative Empirical Analysis

7.1 Introduction

In the previous chapters of this part of the book, we stressed the importance of structural features of renewable technologies and the role these features play in adoption processes. The result obtained in the previous chapters is that strong irreversibilities in the decision-making process to adopt do indeed exist, despite the institutional rules governing the market. Our main interest is now to test these theoretical results empirically. For this purpose, a database is built on the success factors (or barriers) for adoption (or non-adoption) of renewable technologies in different European cities.

With regard to the technologies, a wide spectrum of possible technological solutions exists on which an empirical analysis could in principle be run. Table 7.1 classifies renewable technologies by sources of energy used by these technologies and groups of final users. Among these, the choice for the technologies to be analysed in our empirical analysis is driven by our interest to cover most of the categories, in order to capture the behaviour of adopters in view of:

a) technologies allowing *energy efficiency improvements*, which increase the productivity of primary energy sources;
b) technologies based on *combined heat and power generation*, which allows a more efficient conversion of primary sources into usable energy forms (CHP);

c) technologies based on *renewable energy sources*; i.e. those energy sources which may be reproduced, and therefore are not subject to depletion. For these reasons, the technologies chosen are: photovoltaic systems connected to an electricity grid; solar-water heating systems; non-individual wind energy systems, energy from refuse incineration; energy from the recovering of biogas from refuse; combined heat and power generation with district heating systems; district heating fed by renewable sources; insulation technologies for savings in energy use.

When it comes to the choice of cities, the aim of the empirical analysis is to cover large and medium/small cities, some of them highly specialized and others with a diversified industrial structure, and a (manifest or latent) interest in renewable technologies, expressed in terms of planned and/or implemented energy policies and projects. For this reason, as mentioned in Chapter 3, four cities were chosen in three different European countries, namely Milan, Turin, Brescia and Pavia in Italy; Amsterdam, Rotterdam, Almere and Gouda in The Netherlands; and Thessaloniki, Chania, Mytilini and Ptolemaida in Greece.

As presented in Section 3.5 of this volume, the sample of cities covers both large and medium size cities characterized by both a diversified and a specialized urban system. For each of them, four different technologies are analysed, obtaining at the end a database composed of 60 observations for each issue[1].

The main research questions stemming from our theoretical model that we would like to test in our empirical analysis are the following:

- do the main structural features of renewable technologies, i.e. indivisibilities, complementary assets, adjustment and sunk costs, play a role in the definition of the perceived advantages of these technologies? In other words, do these features represent barriers for the development of renewable technologies, as expected from our theoretical model?
- do the main structural features of renewable technologies play a role in the future willingness expressed by our city sample, to adopt, as suggested in our theoretical model?
- do the main structural features of renewable technologies play a role in the definition of the optimal time of adoption, i.e. in the identification of an irreversible choice of non-adoption?

[1] The 60 observations for each question are the results of: 3 countries; 4 cities in each country; 5 technologies in each city.

Table 7.1. Renewable Technologies by Energy Sources and Final Users

Sources **Groups of final users**	**Solar**	**Biomass**	**Water**	**Geothermal**	**Energy saving**
Domestic • heating/cooling • hot water production • lighting • electric appliances (compelled electric uses)	• passive solar • solar water heaters • medium temperature solar systems • photovoltaic	• energy waste from combustion to be • used in DH systems • anaerobic digestion for biogas as fuel	• hydroelectric • hydroelectric	• geothermal energy to be used in DH • systems with • methane	• thermal insulation (walls, windowpanes) • C.H.P. and DH (also with non renewable e energies)
Commercial • heating/cooling • lighting • electric appliances • (compelled electric uses)	• passive solar • solar water heaters • medium temperature solar systems • photovoltaic	• energy waste from combustion to be • used in DH systems • anaerobic digestion • for biogas as fuel		• geothermal energy • to be used in DH • systems with methane	
Urban transport system	• photovoltaic				• high efficiency engines
Urban waste management	• solar water heaters (under 100°)	• energy waste from combustion to be used in DH systems • methane from disposals to be used in power plants • anaerobic digestion for biogas as fuel			
Industry		• energy waste from combustion to be used in DH systems • methane from disposals to be used in power plants • anaerobic digestion for biogas as fuel	• hydroelectric	• geothermal energy to be used in DH systems with methane	• C.H.P. and DH(also with non renew-able energies) • micro C.H.P.

Moreover, as suggested in Chapter 4, often too much emphasis is put on the technological aspects in terms of their role in the definition of rhythm and speed of adoption processes of renewable energy systems. Our empirical exercise aims also to test the question: do other bottlenecks than the mere technological ones play a role in the definition of both the advantages achieved and the willingness to adopt in the future?

In this chapter we present the results of our interpretative analysis. In Section 7.2 we present the methodology and the indicators used in the analysis. Sections 7.3 and 7.4 provide the econometric results of our interpretative model. Section 7.5 contains some concluding remarks.

7.2 Methodology and Indicators

The methodology used in our analysis is based on a discrete choice modelling approach, in particular a multinomial logit model. The choice of this methodology is driven by the essentially discrete nature of our database on the choice of a city to adopt these technologies. On the basis of a series of explanatory variables, representing the main structural features of renewable technologies, a series of logit models is estimated to explain:

- the willingness to adopt these technologies in the near future;
- the perceived advantages generated by the adoption of these technologies;
- the optimal adoption time.

The explanatory variables for the interpretation of these three elements are grouped under two different frameworks:

- the *techno-economic features* of renewable energy systems, i.e. whether they belong to the so called "entry barriers" (indivisibilities and complementary assets), or to "exit barriers" (sunk and adjustment costs);
- *the different nature* of the techno-economic features, i.e. whether they are of an organizational, economic, financial, ecological or technological nature. This distinction is based on our concept of a pentagon of concerns, presented already in Chapter 3; our approach emphasizes the importance of different

barriers to adoption, which go beyond the technological aspects, which are too often conceived of as the main driving forces for the diffusion.

Table 7.2 groups the critical factors emphasised by our survey questionnaire under the two above mentioned frameworks. In the rows the different kinds of techno-economic features are presented, while their nature of organizational - rather than technological, market, financial and ecological - barriers to adoption are shown in the columns. For each of them, Table 7.2 reports the different potential critical factors of adoption which are mentioned in the questionnaire. Interviewed people were asked to express a judgement on the degree of importance of the different critical factors in the development of a particular technology. The sample of people to be interviewed was chosen among local policy-makers, suppliers and experts in renewable technologies in the twelve different cities under investigation.

Indicators are built next in order to obtain a synthetic measure for each column and each row: the modal value is chosen as a synthetic proxy for all different possible replies belonging to each column and each row. In this way we were able to obtain:

- an indicator concerning *indivisibilities* of a different nature: technological, economic, financial and ecological/institutional indivisibilities. In this indicator critical factors like a critical mass of users and of know-how, a minimum size of plants, an initial financing level, and a critical level of organizational cooperation among users, are contained;
- an indicator concerning *complementary assets* of a different nature: technological, economic and ecological/institutional complementary assets. In this indicator, critical aspects like technological compatibility between old and new technologies, a critical level of technical know-how of suppliers of complementary technologies, and retrofit problems of already existing buildings, are contained;
- an indicator concerning *adjustment costs* of a different nature: organizational, economic and ecological/institutional adjustment costs. Requalification of staff and of management functions, changes in the mentality of users and suppliers, and awareness of local citizens are all critical factors present in this indicator;
- an indicator concerning *sunk costs*, typical of an economic nature. This indicator contains in fact only losses due to incomplete depreciation of the capital stock on the old technologies, and sunk costs on the new ones;

- an indicator concerning the *advantages* of renewable technologies. This indicator contains:
  a) economic advantages, like economic efficiency achieved with the new technologies, suppliers revenues, profitability of users;
  b) technological advantages, like technical trustworthiness of the system for final users, energy advantages obtained with respect to initial targets, and
  c) ecological advantages, like reduction of air pollution and green house effect thanks to the introduction of new technologies.

A difference to the theoretical model presented in the previous chapters is that the empirical analysis is run while taking into account both energy suppliers (potential adopters) and final users; in this respect, for example, adoption advantages do not relate only to revenues of energy suppliers, like in the theoretical model, but encompass also the advantages for final users.

Three additional indicators[2] are added to the categorical variables (Table 7.2). These are:

- an indicator concerning the *external conditions* which could influence the adoption choice. The degree of freedom of the local energy suppliers in the decision to innovate, and the climatic conditions are an example in this respect;
- an indicator for *the optimal time of adoption*, by analysing the judgement of interviewed people on the present time as the most appropriate time for adoption. The aim is to measure the importance of the adoption time in the decision-making process. In particular, interviewed people were asked whether the present time was the most appropriate moment for the adoption to take place;
- a *country* variable (dummy), discriminating between Mediterranean countries, i.e. Italy and Greece, and Northern Countries, in particular The Netherlands. We would like to capture in this way whether some country effects exist in the decision to adopt.

Moreover, another group of explanatory variables was built, with the idea of testing whether the main bottlenecks and barriers to the willingness to adopt

[2] Two other variables were used; a variable distinguishing between small/medium and large cities, and a variable distinguishing between technologies for energy use improvements and technologies based on new energy sources. Both variables turned out to be not statistically significant.

renewable technologies in the near future are of organizational, economic, technological, ecological/institutional and/or financial nature. For this reason, synthetic indicators were built for our pentagon of concerns[3], namely:

- an indicator for *technological barriers* to adoption (called in our pentagon model the "hardware");
- an indicator for *economic barriers* to adoption (the "software");
- an indicator for *ecological/institutional barriers* to adoption (the "ecoware");
- an indicator for *financial barriers* to adoption (the "finware");
- an indicator for *organizational barriers* to adoption (the "orgware").

As before, the modal values of all possible critical factors concerning technological, economic, organizational, financial and ecological/insitutional bottlenecks to adoption are chosen, and used as synthetic indicators for each column of Table 7.2. In this summing-up procedure, the row related to the "advantages" has been ignored, being the indicators obtained by synthetic measures of barriers to adoption.

Table 7.3 covers the interpretation given to each indicator. In the case of techno-economic features these assume a value of 0, if they represent a success factor; if they represent a bottleneck for future adoption, they assume a value of 1. For the 'pentagon of concerns' approach, a value of 0 applies if the variable represents a bottleneck, while a value of 1 applies if they represent a success factor. The three dependent variables for our models, assume that:

- for the advantages and willingness to adopt, a value of 0 applies if they are low, and a value of 1 if they are high;
- for the evaluation regarding the appropriate moment in time to adopt, a value of 0 applies if the evaluation is positive, and a value of 1, if it is negative.

---

[3] With respect to the pentagon of concerns presented in this case, the "software" component is mainly related to flexible adjustments to economic/market elements. In Part C of this volume, the "software" component is more used to indicate information provision and communications to citizens in order to induce environment-benign behaviours. This difference is due to the different aims for which the pentagon model is used: in the present chapter it is related to the evaluation of the techno-economic features and of their different natures, while in Part C the aim of the empirical analysis is the evaluation of urban energy policies.

Next, three logit models were run between three successive distinct dependent variables and the techno-economic features presented above. In the first model, the aim is to explain the perceived advantages obtained from the adopted new technologies. We would expect to find a negative correlation between the advantages of the adoption and the presence of high barriers represented by indivisibilities, complementary assets, adjustment and sunk costs.

The second model aims to describe the role techno-economic features play in the willingness to adopt these technologies in general in the near future. We would expect, from the results obtained in our theoretical exercise, that the willingness to adopt is negatively correlated with some of the techno-economic features of new technologies.

In the third model, the aim is to test the importance of these techno-economic features on the appropriate adoption time. We would expect that techno-economic features play a negative role in the definition of the most appropriate adoption time. In view of indivisibilities, complementary assets, sunk and adjustment costs, the inclination is to postpone the choice of adoption, and thus define the present time as an inappropriate time of adoption.

Another group of models is estimated, with the aim to test one of our conceptual ideas, i.e. whether the nature of the adoption bottlenecks of these technologies are not only (and probably not at all) caused by technological barriers. As argued in Chapter 4, the phase of development in which these technologies develop is the first - and less complicated - phase of adoption, where the integration level with the national ditribution is very limited. The step towards the second phase, that of a more integrated system within the more widespread national energy supply system, requires to overcome some market indivisibilities. From the technological point of view, no barriers exist thanks to the mature phase in which both the core technology and the complementary technological systems exist. The aim of our statistical exercise is to see whether the technological aspect really plays a minor role in the definition of both the present advantages stemming from the new technologies, and especially the willingness to adopt these technologies in the near future.

Before running the logit models, a selection of all explanatory variables is made, based on a selection of a large number of potential explanatory variables. Among all possible categorical variables, those having a greater dependency on dependent variables, are chosen (Table 7.4). In the next Sections, the results of our estimated logit models are presented. Section 7.3 contains the results of the model where categorical variables are the techno-economic features presented above. Finally, Section 7.4 contains the results for the pentagon of concerns approach.

Table 7.2. Taxonomy of success factors in renewable energy adoption process

	Organizational aspects	Technological aspects	Economic aspects	Financial aspects	Ecological/Institutional aspects	
Indivisibilities		• Achievement of a minimal size of plants	• Critical mass of R&D • Technological know-how of suppliers • Critical mass of small users • Critical mass of know-how	• Initial public financement • A certain level of financial participation of users	• Level of organizational cooperation among users, suppliers and local institutions • Presence of an ecological awareness among potential users	Entry barriers
Complementary assets		• Retrofit problems of buildings where the new plant is installed • Technological compatibility between the new and the old technology	• Level of technological know -how of suppliers of complementary assets		• Visual impact of plants in the local environment in which it is located	
Adjustment costs	• Adjustment costs in the management functions • Requalification of existing technical staff		• - Adjustment costs for small users to the initiative • - Adjustment costs for large users to the initiatives		• Awareness of local citizens on the new initiatives	Exit barriers
Sunk costs		• Losses due to the incomplete depreciation of existing capital stock • Sunk costs of the new technology	•		• - Harmony with possible public quarter functions	
Advantages of renewable technologies		• Technical trustworthiness of the system for final users • Energy production obtained with respect to initial targets	• Economic efficiency with respect to expectations • Suppliers revenues • Profitability of small users • Profitability of large users		• Presence of intangible collective benefits in the adoption of the new technology • Reduction of air pollution and green house effect thanks to the introduction of the new technologies, with respect to initial expectations	
External conditions	• Role of a particular technology within the supply of the energy in cities • Market conditions in the energy industry (degree of freedom of energy suppliers) • Visual impact of the new technology • Favourable climatic conditions for the use of particular technology					
Irreversibilities	• Appropriateness of the time adoption of new technologies					

Table 7.3. Definition of variables

VARIABLES	0	1
Categorical variables:		
Indivisibilities	success factor	barrier
Adjustment costs	success factor	barrier
Sunk costs	success factor	barrier
Complementary assets	success factor	barrier
Technological aspects	barrier	success factor
Organizationa aspects	barrier	success factor
Ecological/Inst. aspects	barrier	success factor
Financial aspects	barrier	success factor
Economic aspects	barrier	success factor
External conditions	success factor	obstacle to adoption
Country	The Netherlands	Italy and Greece
Dependent variables:		
Advantages	Low adoption advantages	High adoption advantages
Optimal time of adoption	Positive judgement on the present time as a right moment of adoption	Negative judgement on the present time as the most appropriate moment of adoption
Willingness to adopt at a large scale	Low probability of future adoption	High probability of future adoption

Table 7.4. Correlation matrix between the dependent variables and the categorical variables

Dependent variables / Categorical variables	Advantages	Willingness to adopt	Appropriate time of adoption
Adjustment costs	0.020	0.864	0.24
Complementary assets	0.966	0.409	0.34
Indivisibilities	0.008	0.049	0.88
Appropriate time of adoption	0.242	0.367	-
Sunk costs	0.001	0.561	0.04
Advantages	-	0.11	0.24
Countries	0.000	0.072	0.23
External conditions	0.053	0.128	0.21

7.3 Results of the Estimated Logit Models: the Role of the Techno-economic Features

7.3.1 The determinants of the advantages of adoption

The first aim of the empirical analysis is to find out the determinants of the advantages of adoption achieved so far in the twelve cities analysed. In particular, the interest lies in the definition of the role played by indivisibilities, complementary assets, adjustment and sunk costs in the definition of the advantages of adoption. Our expectation is that these techno-economic features are negatively related to the advantages of adoption: the higher the indivisibilities are, the lower the advantages, since higher indivisibilities inevitably mean higher adoption costs, lowering the profitability for adopters.

The results of the estimated model are presented in Table 7.5, and they are satisfactory from both a statistical and an economic perspective. First of all, the model has a good statistical significance in general, with a Chi-square of 33.94, and a favourable significance level (see Table 7.5). There are four main categorical variables explaining the advantages of adoption. Indivisibilities and sunk costs are both statistically significant variables (expressed by a Wald test of, respectively, 5.82 and 3.11), and are negatively correlated with the dependent variable. This result confirms our expectations: higher indivisibilities and sunk costs decrease the advantages achieved by adopters.

Another explanatory variable is the country variable: it shows that Italy and Greece have achieved advantages from the adoption of renewable technologies, while in The Netherlands this view is not present. There can be many interpretations of this result: it may express a more critical approach to new technologies by Dutch policy-makers, but it could also express that, since The Netherlands in general has a lower penetration level, as witnessed in Table 4.1 of Chapter 4, it is not able to express a positive judgement from the results achieved so far.

Another significant variable is formed by the external conditions. They are statistically signficant and negatively correlated with the advantages of adoption: this means that if external conditions are not favourable, the advantages of adoption are lower. This result stresses two important aspects:

- institutional rules - and especially the possibility of local suppliers to be free to decide on the adoption - are important for the success of the adoption;

- general conditions, like favourable climatic aspects, play a role in the adoption process, as reasonably expected.

Table 7.5. Results of the estimated logit model for the advantages of renewable technologies

Variables	B	S.E.	Wald
Constant	2.66	0.83	10.31
Country	-2.97	0.86	11.69
External conditions	-1.62	0.87	3.42
Indivisibilities	-2.25	0.93	5.82
Sunk costs	-1.36	0.77	3.11

Chi-square: 33.94
df Significance: 0.0000

7.3.2 The determinants of the willingness to adopt in the future

In the previous sub-section the determinants of present advantages have been identified; in the present sub-section we deal with the determinants of future adoption. In particular, one of our main aims in the empirical analysis is to define the determinants of the *willingness to adopt these technologies in the near future*. Our expectation in this respect is that the willingness to adopt is highly dependent on the techno-economic features of renewable technologies, which negatively influence their future development and market penetration.

The results obtained from our statistical analysis are presented in Table 7.6, which contains both the estimated coefficients and the statistically significant role played by each categorical variable measured with the Wald test. The adoption advantages stemming from these technologies seem to play an important role in the definition of the willingness to adopt. This confirms the idea, expressed already by other authors (OECD, 1995), about the importance of success stories of previous adoption for the development of these technologies. As has been advocated for some particular technologies, like CHP, the public perception is often based upon previous bad experiences with poorly designed and constructed systems: lack of consumer heating control, no metering and unreliable systems resulting from delays in repairs and maintenance, have given CHP a bad reputation, which influences the willingness to adopt in the future. Low success in previous adoption co-determine the decision to adopt in the future.

Table 7.6. Results of the estimated logit model for the willingness to adopt renewable technologies at a large scale in the future

Variables	B	S.E.	Wald
Constant	-1.17	0.85	1.88
Country	2.34	0.88	7.03
Adoption Advantages	2.00	0.87	5.25
Indivisibilities	-0.80	0.61	1.80

Chi-square: 14.54
df Significance: 0.0023

Moreover, among all possible categorical variables expressing the techno-economic features of these technologies, indivisibilities are negatively linked to the willingness to adopt. Our main assumption is thus confirmed: in the presence of high indivisibilities, the willingness to adopt in the future is inevitably lower. This also confirms our theoretical model on the three phases of adoption, presented in Chapter 4. The move from the stand-alone phase to the grid integration phase, which is the one our interviewed cities are going through at the moment[4], is in theory expected to be hampered by indivisibilities of different kinds: the achievement of a critical mass of users, of a mimimum size of plants, to exploit economies of scale, a critical level of campaigns in favour of renewable technologies. Our empirical results seem to confirm these expectations.

A last categorical variable seems to be the country variable: in particular, the willingness to adopt is higher in The Netherlands rather than in Italy and Greece.

[4] This assumption is based on the qualitative judegement expressed by our experts during the interviews.

There may be several explanations for this result: more awareness of more advanced countries of the importance of environmental protection; more public finances dedicated to local environmental projects; or more awareness of local policy makers on the importance of environmental quality for economic efficiency of local systems and for local competitiveness. But it may even stem from the lower penetration level achieved by these technologies in The Netherlands, and by their awareness that since low advantages have so far been achieved with these technologies, other more positive results can be obtained by their exploitation. Not surprisingly, this variable plays the opposite role, when used to explain either the present adoption advantages, or the future willingness to adopt. The country with the perception of achieving low advantages from adoption is the one which has higher interests in adopting in the future, aware that more positive results can be achieved via the introduction of these technologies.

The empirical results of this part of the analysis confirm in any case our expectations: *the presence of indivisibilities in renewable technology systems negatively influences the willingness to adopt*, as suggested in our theoretical model.

7.3.3 The determinants of the optimal time of adoption

In our theoretical model, much emphasis is given to adjustment and sunk costs, complementary assets and indivisibilities for the definition of the optimal time of adoption. Our main interest in this part of the analysis is to confirm empirically this theoretical result, which stresses the existence of irreversibilities in decision-making processes.

For this purpose a particular logit model is estimated, where the dependent variable is the perception of interviewed people on the present time as the most appropriate moment to adopt. The results of this model are presented in Table 7.7. The categorical variable which is statistically significant to explain the appropriate adoption time is the indicator of sunk costs. In particular, the results of our analysis show that the moment of adoption is less appropriate, the higher the sunk costs are. Also in this context, our expectations are satisfied, mainly achieved thanks to only one significant variable, since other possible categorical variables were, unfortunately, not statistically significant.

Table 7.7. Results of the estimated logit model for the definition of the appropriate time of adoption

Variables	B	S.E.	Wald
Constant	-0.77	0.49	2.45
Sunk costs	1.20	0.59	4.07

Chi-square: 4.32
df Significance: 0.037

7.4 Results of the Estimated Logit Models: an Approach in Terms of the Pentagon of Concerns

Another approach to the analysis of the determinants of both the adoption advantages and the willingness to adopt in the near future is the one based on our pentagon of concerns. In synthesis, our main approach to the barriers and bottlenecks of the adoption of renewable technologies is based on the assumption that the barriers are not only of a technological nature. The development of renewable technologies is hampered by organizational, financial, market and ecological/institutional elements which play an important role and which - in the explanation of renewable technology adoption processes - go well beyond the simple technological problems.

To analyse the role of barriers of another nature than the mere technological aspects, synthetic indicators of organizational, financial, market and ecological success factors in the decision making process to adopt, are analysed. In particular, the aim in this part of the analysis is to investigate:

- whether the success factors determining the advantages of adoption are of a different nature;
- whether the willingness to adopt in the near future is dependent on technological upgradings of renewable energy systems, or also on other factors, like oganizational, ecological/institutional, financial and market changes.

The results of the advantages of adoption are presented in Table 7.8. In the analysis, there are two statistically significant categorical variables which explain the advantages of adoption. On one hand, the country variable is always very significant as in the case of the analysis run in the previous Section, where the advantages of adoption are more clearly perceivable in Italy and Greece, than in The Netherlands. On the other hand, the importance of economic variables in the determination of the adoption advantages is, evident from the results. The major result of this analysis confirms our expectations: *the advantages of the adoption of renewable technologies do not depend on technological aspects, but more on economic success factors*. The absence of these economic factors represents a real bottleneck for achieving significant advantages. This view is in line with the interpretation of the degree of market penetration as the main obstacle to the achievement of significant advantages from these technologies.

Table 7.8. Results of the estimated logit model for the pentagon model on the adoption advantages

Variable	B	S.E.	Wald
Constant	0.35	0.63	0.31
Financial success factor	0.79	0.68	1.35
Economic success factor	1.30	0.79	2.88
Ecol./Inst.success factors	1.08	0.81	1.75
Technological success factors	-0.55	0.82	0.45
Country	-2.34	0.74	9.99

Chi-square: 23.66
df Significance: 0.0001

Moreover, another quite significant variable identified in our model is the so-called ecological/institutional variable; when these aspects represent a success factor, the advantages achieved from these new technologies are greater. This has two important consequences:

- the institutional rules governing the industry are extremely important in the definition of the advantages;
- the ecological aspects of renewable energy technologies, like their visual impact, as well as campains to promote citizens' awareness of the importance of these technologies for the environment should be explicitly kept in mind.

Table 7.9 contains the results of the success factors as explanatory variables for the willingness to adopt these technologies in the future; also in this case, the results confirm our expectations. The willingness to adopt *these technologies does not depend on technological elements*, but more on *economic and financial elements*. The role played by these success factors on the willingness to adopt is statistically significant, while this is not the case for the technological elements.

Table 7.9. Results of the estimated logit model for the pentagon model on the willingness to adopt

Variable	B	S.E.	Wald
Constant	1.68	0.77	4.77
Financial success factor	1.22	0.77	2.52
Economic sucess factor	1.03	0.68	2.26
Ecol./Inst. success factor	0.28	0.75	0.13
Organizational success factor	0.51	0.72	0.50
Technological success factor	-0.21	0.73	0.08
Country	-1.54	0.77	4.00

Chi-square: 10.83
df significance: 0.09

As a general result of this part of the empirical analysis, we underline our general impression: barriers of another nature than the mere technological ones are nowadays influencing the rhythm of adoption of these technologies. In the conceptual approach of the three different phases of adoption presented in Chapter 4, our results witness that the economic and financial barriers are at present very important to achieve the second phase, i.e. that of national coordination and integration of small renewable systems. As we will point out in the part of the book dedicated to local policies, strategies in favour of the development of technological aspects are not sufficient: technological barriers seem, in this phase, to be less stringent than financial elements and economic aspects, which instead appear to play a more crucial role in this phase of adoption and development.

7.5 Conclusions

Some interesting results have emerged from this first empirical exercise, which are in line with the theoretical assumptions developed in previous chapters. In particular, the main statistical findings may be synthesized as follows:

- the advantages achieved through the adoption of the renewable technologies depend on the presence of techno-economic features which still hamper the full exploitation of new technologies. In particular, the presence of indivisibilities and sunk costs hamper the achievement of the advantages, being negatively correlated with the dependent variable;
- the willingness to adopt in the near future is explained by two important elements:

  a) the presence of success stories of previous adoptions (being positively correlated with the advantages achieved so far), and

  b) the presence of indivisibilities which inhibit the willingness to adopt, and thus support the interest for a postponed adoption;
- the appropriate time of adoption is determined by the presence of sunk costs: the higher the sunk costs, the less appropriate the present adoption time. This also means that in the presence of high sunk costs, potential adopters are more willing to postpone the adoption, in order to achieve full depreciation of existing capital;
- technological elements do not represent the main reasons for explaining the low adoption and the low willingness to adopt. In both cases, elements of different nature, and in particular of an economic and financial nature, explain

both the success obtained by past adoptions and the willingness to adopt in the future.

From the empirical analysis an interesting result emerges: both the technology and the city variable turn out not to be statistically significant. In this respect, this seems to suggest that neither technology nor the specific peculiarities of the city play a crucial role in the adoption process, at least not in the aggregate empirical analysis carried out here.

The previous empirical analysis is related to the theoretical model on the optimal adoption time developed in the previous chapters. However, the database built from our empirical analysis contains more information on the diffusion of different technologies in different urban contexts. In the next part of the book (Part C), a more in depth analysis, both quantitative and qualitative, is undertaken with the aim to evaluate urban policies on renewable technologies, incorporated in our city sample. The evaluation is developed by taking into account the reasons for adoption, the degree of penetration of the different technologies and the main barriers for a widespread diffusion of these technologies in different urban contexts. In the last part of the book policy implications and normative suggestions are highlighted.

PART C

COMPARATIVE ANALYSIS OF EUROPEAN SUSTAINABLE CITY OPPORTUNITIES

8 Success Factors and Barriers of Urban Energy Policies: a Qualitative Analysis

8.1 Introduction

In this chapter we will analyse qualitatively various relevant aspects of the Pentagon model of critical success factors. The information source for our analysis is based on information offered by various qualified experts involved in the energy initiatives of the above case cities. They comprised amongst others policy-makers, engineers and representatives from energy companies.

This chapter provides an *exploratory and explanatory* analysis of the weakness and strength profiles of the various urban renewable energy technologies, by mapping out the type and relevance of the problems associated with these aspects, as well as the causes of these problems, for all the critical aspects specified in Table 3.1. In the subsequent Chapter 9, a comparison will be presented of the perceived order of *magnitude* of bottleneck factors of renewable energy policies among various technologies, based on the same series of interviews.

8.2 Technological Critical Factors

In this section we will address the technological factors that are decisive for the introduction and adoption of renewable energy initiatives in cities while in a subsequent section (Section 8.3) we will address in more detail market-related factors. In the present section we will focus on R&D, technological know-how, climatic conditions, the role and size of technology and technological compatibility, respectively.

8.2.1 R&D activities

R&D is one of the prominent vehicles for stimulating technological progress in renewable energy systems. The intensity of research activities on district heating systems has, in many countries, shown a rapid pace, although achieving it has required many efforts and expenses. It seems likely that the more advanced forms of district heating, which have been implemented on a much smaller scale (e.g. applications using waste as energy source), have to face in the future high R&D costs to achieve a significant improvement, especially with regard to their environmental impact, and to lower the costs of incineration plants that presently weigh heavily on the overall investment costs. Clearly, the situation differs in the case of PV systems (and to a minor extent for aeolian systems); research in this field is not only very costly, but it has also been idle for many years.

8.2.2 Availability of technological know-how and technical competences

Various technologies in the field of renewable energy sources and innovative energy production processes appear to be easily available, even if they (usually) stem from other European countries where the exploitation of this kind of initiative has been at a higher level in the course of the years. Anyway, the reliance on foreign technologies, when a sufficient national support system does not exist, is not perceived as a negative element, because international scientific exchanges are not difficult and the possibility of learning due to foreign experiences are considered by most of the interviewed people as a success factor[1].

Here it is essential to make here a distinction between those renewable technologies that are internationally already in an advanced stage of development

[1] When the technological problems have an international dimension, the situation of course changes. This is the case of PV systems: the technology is very simple, but a larger effort in R&D would be necessary at an international level in order to lower costs which are at present too high. The level of R&D and technological know-how in effect is very low. The research is not completely independent and the market is not competitive because of monopolistic competition. *Indivisibilities* in R&D activities then represent a strong bottleneck for the development of this kind of initiative.

and implementation, and those technologies that are more recently being, because of the restrictive role of the resources to be spent on the acquisition of technological know-how. In the first case, relatively more difficulties will be met at the beginning of the adoption processes, because even abroad, its availability is less assured and a lot of efforts have to be made to achieve a sufficient level of know-how for the local application itself. Paradoxically, in the second case, thc late introduction of new technologies offers more opportunities due to the high level of technological knowledge achieved in the international scientific field. Less efforts in the field of R&D have then to be made and the learning costs can be kept rather low.

This phenomenon has, of course, a counterpart, i.e. the difficulty of achieving a good level of *technical competence*, because it is not easy to follow the progress originating from the outside and often there are problems of technical specialization. The reliance on a third party with more technical competence is not always a positive solution, especially as far as the less implemented technologies are concerned (e.g. waste as renewable source); sometimes there are not many actors who operate in this sector and the choice cannot be free, but can only be made by following the few existing operators who do not necessarily have a proper technical background.

The second possibility is to rely on the re-qualification of human resources, which is accompanied by high costs. This happened in the past and at present thus does not represent a problem in the case of more applied technologies (like traditional DH), but the situation differs for those technologies with an as yet unconsolidated development[2].

The reason of this weak position stems from an important consideration that necessarily involves other fields to be investigated. The question is whether in cases where a 'historic' separation between management of waste and energy production exists, a technical re-qualification would be possible and not too costly. This question is important, because management of refuse as an energy source needs particular technical competence, to take into account the variable composition of materials contained in the waste (which means, among other things, a less reliable

---

[2] These considerations are also valid in the case of other technologies connected to the primary one: sometimes - and always for the less consolidated ones - various difficulties exist to find a secondary technological context in which the specific initiative can easily operate. It is the case of the incineration of waste to recover energy that operates in a sometimes contradictory context of waste management (where an unclear and inefficient destination of different kinds of refuse is typical) or the case of PV systems (simple but technologically costly) that meet difficulties, as already noted, also with regard to the related technologies (e.g. good inverters are often not available).

performance and stronger than usual safety parameters). Clearly, the actors involved with waste problems and also with energy production have less technical difficulties when the challenge is to face the production of energy by using the incineration of refuse as a source.

This problem is of course related to the wider issue of the organizational integration of the urban energy system. It will become more clear in the following section why the upstream integration usually emerges as the best way to deal with energy problems.

8.2.3 Bioclimatic conditions

Bioclimatic conditions are a primary basis on which any renewable and energy savings initiatives rely. This basis has two distinct levels. First, the *source* of energy is strongly related to the prevailing bioclimatic conditions. Intensities of wind, solar radiation, geothermal gas etc. depend exclusively on the climatic characteristics of a region. Second, the *needs* for energy and its forms differ according to the climatic setting. Indeed, they are different for northern and southern European countries[3].

Another consideration concerns the impact of climatic conditions on the development of various renewable energy initiatives. In general, these conditions do not affect the success of the initiatives, once they have been introduced. Different of course is the case where the lack of continuous sunshine and the presence of foggy weather (e.g. in the cases of the Italian cities of Milan and Pavia) make the introduction of PV systems unlikely, or where the lack of a sufficient wind speed is prohibitive for the use of aeolian systems.

In general, it may be remarked that particular climatic conditions can sometimes affect the *technological features* of the plants, with costly technical and environmental consequences. This is e.g. the case for the smoke chimney of incinerators: humidity, windiness and atmospheric pressure affect the structure and the construction of the plants. When the waste chimney is high enough, it calls for a

---

[3] In the Greek case the climatic conditions are in favour of renewable energy. Specifically, the particular characteristics of the urban sites (case cities) determine to a large extent the appropriate energy form. In all successful projects these characteristics were very promising for the establishment of the relevant technologies. In this respect, the location of the projects was a decisive factor for their success and were taken on the basis of strict scientific criteria. On the other hand, the demand for energy, its specific characteristics and its allocation in time, have to play a crucial role. They determine to a considerable extent the energy production pattern and therefore the success of the projects.

wide 'fallout cone' and a low concentration of pollutants, but when the waste chimney cannot be higher than 35 meter, the technical costs will be higher and noisy blowing machines become necessary.

A morphological bio-climatic factor that has a strong importance is the appropriate of the possible spontaneous natural sources that sometimes can be found: for instance natural heat sources from the earth or the presence of large quantities of water. In the cases investigated in our empirical study however, another factor similar to a spontaneous source, seems more critical and worth attention, viz. the excess of refuse to be sold off, or biogas from disposals that in many contexts becomes very problematic and serious due to the absence of space for additional disposals, and in particular due to the opposition of the population to the existing ones.

The way to get out of this situation caused by the excess of refuse may be to emphasize the role of refuse and biogas just as a renewable source to be burnt to produce heat and/or electric energy. Unfortunately, this challenge has not been met everywhere and the gap is still great between waste overproduction and its optimal re-use in the context of integrated waste management that minimizes the reliance on the disposal and maximizes the waste value.

8.2.4 Role of the technology

In general, it can be stated, for all our investigated initiatives, that the strategic role of the technologies can emerge *independently* of the quantity of renewable energy actually produced. This is valid, for example, for those technologies that regard the use of biomass. It is clear that they cannot at present supply to the cities a large quantity of energy and the dependence rate of the city on these initiatives is low. However, they represent a very important element of the integrated waste management that should not be undervalued, and sometimes the level of energy produced is in line with the city objectives regarding the energy production from renewable sources.

It is true however, that in the case of district heating, which at present may be regarded as the most important technology to save large quantities of energy, the larger its scale can be, the more important its role becomes. This happens especially (as we see in Section 8.6) when its supply is *integrated* with the methane supply.

Finally, these considerations are not valid in the case of PV energy. These systems are still in most circumstances exclusively demonstration cases and neither efficient nor profitable. This makes them of little interest to many cities, and on top of that, they produce a low level of energy, at least in our investigated cities.

8.2.5 Efficient size of the technology

It is difficult to speak generally about a minimal plant size in order for renewable technologies to operate efficiently. Often this efficient size depends on contingent conditions but in general a large plants is not considered necessarily as a positive element in achieving an optimal regime of production. In the case of district heating, the large size allows to achieve a great efficiency, but equally important is that such efficiency is achieved gradually. When the policy of cities (and this is, for example, the case for enlargement projects of the district heating systems of Turin, Almere and Rotterdam) aims to move rapidly from a small implementation scale to a larger scale, many technical problems can emerge. In addition, the *adjustment costs* and the *indivisibilities* linked to a necessary large size (which is difficult to be achieved), may become very high. In order to reach such size, rather complicated and expensive technologies are necessary and there may be difficulties in finding external operators to substitute the internal staff to be re-qualified. This may make future returns in terms of profitability of the projects uncertain because of the high costs connected to it.

This is different from case where, also in the long run no efficient size of the technology can be reached. This may happen with large-scale applications in areas with low population densities[4].

In other cases a large size may appear to be very efficient as far as the energetic aspects are concerned, but may be impossible to be achieved because of the high costs involved. This applies also to the case for PV systems, which, if implemented

---

[4] Indeed in some cases, the size of the technology functioned as an obstacle for the early establishment of projects. Some technologies are associated with a high degree of indivisibility and therefore, they require a high level of demand. Such a critical mass of demand could not be created e.g. in Greece due to a relatively low population density. In this respect, the majority of the existing projects in this country is based on relatively medium sized technologies with a flexibility in their divisibility. Nevertheless, it should be stated that the indivisibility of some renewable energy technologies is not a prohibitive obstacle since there is a fairly high number of technologies which could be applied to a small and medium size scale as well.
Over the past years, some bigger projects have been designed, based on large scale technologies at the production level, being at the same time suitable for small individual users spread out over space. It seems that the current maturity of the relevant technologies permits their application in cases where it was previously impossible.

on a large scale, also have a strong impact on the location (because of the absence of proper space). For this reason, a small size is prevalent for existing PV systems, a size that would be efficient only in an isolated configuration.

8.2.6 Technical compatibility

The level of technical compatibility between old or traditional technologies and new technologies planned to produce energy in a more efficient way affects both energy companies *(supply side)* and users *(demand side)*, in the sense that they may face costs in terms of technical re-conversion.

It can be stated that the level of technical compatibility between the old technologies and the new ones is critical and represents a success factor. However, if an incompatibility would occur, this would not necessarily create an insurmountable problem to the innovative process. To support this argument, it is necessary to keep a distinction amongst the different technologies investigated, as will be outlined now.

In the case of systems of district heating the switch from gas oil (diesel), the usual alternative source for heating, does not imply large difficulties for the producers, because the old networks fit the new technology[5] well, and old thermoelectric plants can be used to burn methane or other fuels to produce heat for district heating. Furthermore, there are no changes on the *demand side* in the sense that for the users it is only a matter of switching from individual boilers, or from a centralized system fed by gas oil, to a centralized system fed by DH. This happens *inter alia* when the plants of users are very old, which generally entails an absence of sunk costs for them.

Switching from a system fed by methane to a DH system, would be another matter. This would create major problems as technically it is too difficult. This is the reason why a coexistence between methane and DH would have to be encouraged (see Section 8.6).

A far as PV systems are concerned, the technical compatibility is intrinsic when we consider them as systems connected to the electric network and *integrated* to the electric system (even though some problems may arise from the inverters when a

[5] If this problem is generalized however, the reliance on existing infrastructures can have only a success limited to the initial step of market penetration of renewable systems while technical inefficiencies can emerge on the long run Camagni and Capello 1998; see also Chapter 5).

lack of balance to the network beyond some size may emerge)[6].

When biomass is the energy source, a different consideration is in order. When we consider the option of recovering biogas to produce electric energy, a comparison would have to be made between the simple disposal plant and the disposal plant equipped with recover systems. For this technology it is possible to speak of compatibility, because in general the disposal plants have been built in a way that fits this new technology very well. In addition, if incompatibility exists in the case of biomass (especially if used in district heating systems), this becomes not an engineering but a materials concern. Traditional energy sources are not subjected to changes while the composition of waste is variable. This is not a large problem however, because it only means that the safety parameters have to be more strict. Even when this may affect the technical performance, it means more safety.

8.2.7 Conclusions on technological aspects

The common feature of all technologies investigated in our study is the *delay* in their adoption with respect to other European experiences. Apart from the district heating system of Brescia, the first in Italy, which was introduced in 1972 and grew gradually in size, most other experiences in our study are rather recent, in particular the ones regarding the use of refuse as an energy source.

Even if it is undeniable that being a pioneer in a new initiative (like the case of Brescia) means larger costs in R&D and in the acquisition of technical know-how compared to other countries, it can be stated that the level of technological availability has not been the determinant factor for this delay. The reason has to be found rather, as we will show below, in the lack of political willingness linked to non-favourable conditions in terms of the organization of the energy supply and in the lack of awareness of the opportunities regarding the production of energy and the management of refuse.

A noteworthy point is the *gradual pattern* in the adoption processes as a success factor in the case of large projects of district heating. When the adoption has been carried out late or with small sizes, a rapid process to achieve in the short term a very

[6] In a theoretical way, it is possible to speak of integration even in the case of a domestic PV system that is not connected, but installed in houses with two different circuits: the alternate one for electric power stemming from the network and the continuous one for a PV system to make appliances work, like television sets or refrigerators. This would require low energy use without overloading the plants (which happens often with the traditional electric network system).

large size can create strong technological difficulties (as in the case of the enlargement project of Turin), even if the large size in the case of district heating is an important element in exploiting better the economies of scale and the demand for energy by particular classes of users (see Section 8.4). Furthermore, the enlargement of the size, or in any case of the exploitation level (for example, in the case of the biogas from disposal) may generate sunk costs on materials or appliances that are non perceived at this moment.

Anyway, at present the introduction of a new technology means less effort in the processes of acquisition of technological know-how, because there is the possibility of exploiting technologies already tested by others, but it is necessary to recognize that the demand is too strong with regard to refuse and those processes to produce energy more efficiently. In general, it can then be argued that the technologies are available, but that strong efforts will have to be made, for instance, in technical re-qualification.

Another kind of consideration is, on the contrary, valid for those technologies that do not find a considerable field of application like PV systems. Strong efforts at international level will be necessary first to make the technology less costly, e.g. by working on silicon and finding the way to extract the crystalline one from the natural silicon (i.e. from sand or quartz). Another possibility, furthermore, could be using the amorphous silicon that is more innovative (and inter *alia could* receive some EC financing). Only in this way would it be possible to accept less favourable climatic conditions that now inhibit the development of PV systems.

Taking into account the difficulties to have PV systems installed on a large scale, another way of development could finally be to find the space for the above mentioned innovative coexistence inside domestic uses (see footnote 5).

Finally, another important issue that in the future could represent a success factor and for this reason has to be exploited carefully, is the following. Besides the necessary employment re-qualification stemming from the introduction of new and innovative technologies, a large economic advantage for the industry originating from the use of biomass for energetic purposes and district heating with or without refuse has to be taken into account. There are two possible advantages. First, the possibility to be connected to pre-existent district heating networks could allow new industrial firms that need power and heat, to find it easily thanks to this system, instead of having to build a new plant to produce them themselves (also with high personnel costs). On the other hand, the development on a large scale of technologies of waste exploitation may stimulate firms to use again their own refuse inside their own production structure by burning them to produce (and use) energy or by means of their reuse and recycling.

8.3 User-related Critical Factors

8.3.1 Awareness of application (information costs)

Information costs represent a very important variable for the technologies of district heating and those ones involving the refuse as energy fuel - especially recently - due to a stronger environmental awareness of population.

The efforts in advertising, public conferences and scientific meetings to illustrate to the citizens the various advantages of initiatives that have also some negative environmental disadvantages, have always shown a very positive effect. There is a necessity to create a climate of cooperation and acceptability (this is the typical case of the district heating system of Brescia above all with regard to the project of using waste). On the other hand, undertaking only few efforts in this direction can represent a real problem for the introduction or the implementation of this kind of initiatives.

Advertising campaigns and information activities play a major role when renewable technologies produce negative impacts and costs in terms of visibility and possible new pollution (noise). In this case, information and advertising campaigns have the role to stress the positive outcomes of new technologies, thus alleviating the negative impacts that these technologies inevitably have. It is the case of PV and aeolian systems which produce a strong negative noise and visibility impact in the area where they are located.

Anyway it is true and obvious, that advertising campaigns can have a positive effect only when addressed to citizens directly involved in the technologies that have to be introduced.

8.3.2 Adaptation of users

The adaptation by small or large users depends on some aspects already analysed or analysed below: technical reliability, economic and environmental advantages and financial support. Here we will summarize through one single concept the willingness to change when that is possible for users (i.e. when it is possible to reject or, to the contrary, to accept new initiatives enthusiastically); a willingness seen as *organizational* aptitude to change.

For many technologies investigated it is not correct to speak about an organizational resistance to changes. In most cases, in fact, nothing changes (if users remain connected to the electric network) or there are only few changes (as in the case of district heating), and these changes then improve the users' situation, facilitating the adaptation[7]. In fact, with a district heating system, the maintenance is centralized and in the hands of the public body that manages the system (in this way the users have no more responsibilities), and the maintenance operations can take place less frequently than in other cases, in light of the higher safety of the technology[8].

8.3.3 Technical reliability

The technical reliability for users represents an important issue on the *demand side* to understand when, if and to which extent the users attach importance to the technical aspects when they are directly involved in a switch to a new technology.

In general, it can be stated that the technologies analysed have a strong technical reliability for users. In some cases, they concern only the production of electric energy and the subsequent connection to the network (with evident safety for users who in reality do not have to face any change in their habits).

In other cases (district heating systems), the reliability is higher as a result of the technological progress made in the past years and because of very strong environmental standards. It is clear that in terms of safety for users and population, it is easier to manage a single great boiler by a Local Board rather than to manage many little boilers by private citizens (or one gas oil boiler localized in a block of flats). Furthermore, *flexibility* is a sign of reliability: there is the possibility of using multifuel boilers to minimize the fuel cost, diversify the resources, and then to have a low price dependency (a '*theoretical*' possibility because it is true that in most initiatives regarding district heating systems the only fuel burnt is just the methane, apart from the projects of DH fed by waste).

Finally, some remarks are in order regarding the other technologies, especially when they do not provide electric energy to be connected to the network. In general,

---

[7] However, the behaviour of users of district heating is perceived by the energy body of Turin (AEM) as not completely flexible. The house administrators in fact seem to oppose some resistance when they prefer to switch on the heating at 6 am and switch it off at 7 pm, while AEM would prefer to offer a lower but continuous temperature during 24 hours (it would be more economic for AEM and could be more economic for the users thanks to lower night tariffs).

[8] Of course, this is less valid in the case of district heating fed by refuse.

the reliability would be high for PV systems, but in the case of the aeolian systems the *uncertainty* has to be taken into account. The continuous *availability* of the wind resource can not be assured and then the cost of an 'emergency resource' is important, in order to face a temporary lack of wind. This perhaps might be of little concern to an Electricity Board, but for a self-producer it could be crucial whether he is connected to the network or not.

8.3.4 Mass of adopters

It is important to know whether the supply side of the technologies investigated fit are suitable for a large mass of users and whether a large demand by users would be feasible.

The demand for renewable energy is created by three forces: first, the real market potential which make users ready to participate; second, institutional reasons which create a demand on the basis of socio-political targets and third, by developing environmental and social awareness among the potential users when the circumstances permit it.

It seems that many cases exist where no project takes place - although the conditions would permit it - since there is a considerable lack of potential users. Therefore, the process of promoting the relevant awareness among potential users is very important. Governmental intervention could be helpful in this direction by developing pilot demonstration projects and informing people as well. In this framework the participation of large scale users may be promising; governmental activities may have to play a crucial role since they could form the critical mass of large scale users for these projects[9].

As far as the built environment is concerned, the district heating technologies are suitable and work more efficiently when (especially in high dwelling density areas) they reach a large mass of users. In this case it is possible to exploit through economies of scale all advantages to have a centralized heating system and in

[9] It should be pointed out that in Greece no large-scale user participated in existing projects. This made the importance of the number of the small and medium users extremely high. However, the lack of large scale users did not prohibit the success of the projects. In all successful Greek projects there was still a critical minimum level of demand. Moreover, the fast development of some projects was based on the fairly high adoption by potential users who saw the participation into the project as a social 'must' activity.

particular a cogeneration system utilizable by many different users. On the other hand, many city districts are characterized by a mass of users that becomes optimal, when - besides big quarters with high dwelling density - there are also large scale users like hospitals, universities, commercial or industrial areas. In this case however, there are not only positive examples stemming from the investigated cities: the case of Pavia is the typical negative example[10]. In fact, its district heating remains at present, but hopefully not in the future, a study of a possible project, despite the optimal conditions of the city in terms of potential large users.

Amongst the other technologies, on the contrary, the mass of users is not an important factor because they are already connected to the network, while for those parties who produce electric energy without a connection (self-production), it is impossible to think in terms of a critical mass of network-linked users.

8.4 Financial Critical Factors

8.4.1 Public and private financing

The cost of new energy technologies is the most crucial factor for the development of renewable energy and energy savings projects. However, an important observation has to be made by making a distinction between large-scale and small-scale projects. For large-scale projects the initial costs are high, while the expected benefits are spread out over the distant future. Therefore, the investment costs could form a strong barrier, especially when there is no relevant experience of similar initiatives. In these cases the investment risk is noticeable and functions as an obstacle. In this respect the existence of proper financial schemes is the necessary condition. In particular, since the development of pilot and demonstration projects is a strong success factor for the adoption of new technologies at a large scale, proper

[10] Unfortunately there are cities that have not yet taken this opportunity, even though they have optimal characteristics for the development of a DH system. The case of Pavia is exemplary: with the S. Matteo hospital and the University Pole, this city represents a strong attractive pole for particular people. The fact that these two areas are essentially "non profit" areas generates a great problem for the city which faced a big energy demand for users who do not bring economic resources to the city. Moreover, the DH could be a necessary tool to save energy just in these areas with a very large demand, but the project has been delayed because an investment of 100 billion Italian Lire would be necessary and it is difficult to find a suitable location for the plants.

innovative financial schemes are vital. For most of the existing large projects in the three countries, the financial system was the most crucial factor. The majority has been financed by European programmes and by governmental schemes. It is evident that without this support, they would not have taken place. Nowadays the problem needs new solutions and the development of new financial models based on private initiatives should be promoted, since the production of renewable energy seems to become a beneficial activity despite high initial costs. There are several indications at present that bank consortiums are aiming to participate in such initiatives.

For small-scale applications where the investment costs are lower, sometimes the project costs do not form serious barriers, but only in favourable (climatic) circumstances[11].

That the financial variable represents one of the most critical factors with regard to the initiatives concerning renewable energy sources, can be illustrated by the situation in Italy. Here, besides a financing provided by the law 308/82 (30% in c/c of declared costs) and another one by the law 10/91, a legislative measure decided by the Interdepartmental Committee on Prices (CIP resolution 6/92) exists that ensures proper prices for all subjects producing electric energy and selling it to the National Electric Company (ENEL) (see also Chapter 3). This means a facility of 150 Lit/KWh for the first 8 years stemming directly from ENEL and another one of 100 Lit/KWh stemming forever from the National Equalisation Fund. In some cases however, with this subsidy from the national legislation it is not possible to repay the total initial expenses and to remunerate the initial investments. This is valid for aeolian resources[12] and for PV systems[13] even if additional financing would take place, as planned by the law 10/91, providing for financial incentives up to 75% of the capital cost[14]. This financing has been deferred due to the financial difficulties of

[11] In Greece, for example, in small-scale projects individuals and small enterprise associations have proven to be able to deal with these project costs. In the past, these small scale applications were supported by governmental financial injections. After the first period which functioned as a demonstration basis now they are able to be financed by the users alone.

[12] In this case the index price is Lit. 155,4/KWh until 2001 and Lit. 72/KWh after that date; if there are incentives *ex lege* 10/91, the price can decrease up to Lit. 120/Kwh. This price, compared to the price of electricity from a thermal resource (Lit. 70/Kwh) shows that at present it is too difficult to develop wind technologies (Lorenzoni 1994).

[13] Without account of financial and fiscal loans, the cost of energy would be Lit. 759/Kwh (and it could go up to Lit. 1500/Kwh if users would live in isles). The comparison to the cost of electricity from a thermal resource is worse than in the case of the aeolian resource and it does not leave space to the industrial development of photovoltaic energy.

[14] Furthermore, as far as PV roofs are concerned, regional financing could be used for savings activities in the building sector *ex lege* 10/91 art. 8.

the country and has given a major responsibility to the financial policy of the EC that has helped, in more suitable areas, most Italian installations.

A different remark has to be made for those initiatives that in general can finance themselves. This is the case for district heating (the case of Brescia) and the case for solar water heating (the case of Gouda), where any public financing was not provided for. It is clear that in this case, other favourable conditions occurred (a high inflation rate that lowered the investment loans and high oil prices) besides the particular political willingness (absent elsewhere) to accept the initial uneasiness related to pioneering projects of these technologies in Italy and The Netherlands. Although other initiatives of this kind have a great benefit from the viewpoint of public financing (due to the presence of unfavourable conditions, as we will show in Section 8.6), the high profitability for most district heating initiatives makes self financing more probable.

In general, the possibility of relying on *private* financing tools has to be taken into consideration. The *'third part financing'* is an optimal work-done contract that could be used by public bodies which want to implement an energy saving project, but do not have the necessary capital and perhaps neither the technical capabilities. In this case the investment may be carried out by a third company, that has to define the investment cost, the necessary time and an energy budget. This sometimes happens for small-scale initiatives concerning the use of biomass, mainly as a result of their technical particularities which sometimes require private management.

8.4.2 Indivisibilities and sunk costs

Sunk costs very much depend on the characteristics of each technology and thus any kind of generalization is problematic.

The introduction of the technologies of district heating does usually not imply sunk costs neither at the supply side nor at the demand side. This happens because energy companies often decide to switch to this system by exploiting old thermoelectric plants that can be used to burn methane or other fuels to produce heat for district heating. On the other hand, users too frequently decide to pass to a district heating system when their plants are very old because the costs stemming from the incomplete depreciation of the capital stock decreases over time and the late introduction is more profitable for users. All this means then the absence of sunk costs on the old technology.

We may also find technologies that are not so profitable because the initial investments cannot be reimbursed. This is the case of PV and aeolian systems which present a threefold problem: the initial costs of the photovoltaic modules and the aero-generators[15] are still too high, while both technologies show a strong indivisibility and are not yet profitable. This prohibits a possible development of these initiatives and makes the presence of the unavoidable sunk costs on technology more stringent.

A further remark has to be made about technologies involving energy recovery from waste. On the one hand, there is the field of waste incineration that requires very strong safety parameters and the existing incinerators might be not suitable to be simply reconverted in plant equipped to recover. This could mean a presence of sunk costs due to the incompleted depreciation of the old technology taking into account also the high costs of the more safe incinerators[16] and the appliances for the depuration of smokes stemming from the burning process of waste. On the other hand, there is the use of the biogas stemming from disposal. In this case some sunk costs of the new technology could emerge: the appliances bought for biogas in the future will not be used because of the exhaustion of biogas and because the pumping could be incompletely depreciated.

With regard to the existing morphology of cities, favourable location areas (in this specific case, industrial depressed areas) often constitute a good occasion to develop district heating systems with positive repercussions on some success factors like the reduction of the initial costs of investment, and the cutting down of possible sunk costs of the old technology thanks to the possibility of using again parts of the old plants.

It is for this reason, for instance, that the city of Milan, which appeared to be a city with an excessive predominance of methanization to the detriment of district heating, can be revalued on the basis of the exploitation of the above mentioned old heating plants around the city. This means that the city is trying at present to integrate the district heating with a methanization process despite the predominance of the methane system for this city. This happens in those areas that are most suitable for this purpose.

---

[15] The 'Dora Riparia' Valley in the Piemonte Region, was the only area amongst the cases considered in our work, that was deemed to be suitable, thanks to its wind speed, for studying the introduction of an aeolian system. After a more thorough analysis however, just the costs of aerogenerators and the comparison with the energy cost from a thermo-electric plant with a combined cycle have led to a discontinuation of this project for the moment (Garbati and Galliano, 1993).

[16] About 4 Million Lit./KWh.

8.4.3 Financial participation of users

When looking at the financial participation of users, it is again necessary to make a distinction among different technologies. With regard to those technologies that provide electric energy, and supply this to the electricity network, often the users of these technologies cannot be distinguished from other users connected to the electric network, and in such cases there is not much relevance for their financial participation.

With regard to the district heating technologies the situation is different. Facilities are granted (or provided) to the users of heat due to the high efficiency rates and moreover an initial 'non participation' bonus to the connection costs as a result of the high profitability for the suppliers.

A remark however has to be made about the possibility that new projects of district heating turn out to be more costly than projects carried out in the past. This can happen due to the disappearance of some favourable conditions that occurred in the past and have already been mentioned: a) the presence of higher engineering costs, b) more economic methane and restrictions on electric energy selling, and c) the necessity to use a new source, the refuse, that makes the district heating more costly because of the incinerator.

In these cases the financial participation of users (such as the participation in the connection costs) is necessary. This is not a simple issue since district heating is less profitable than other energy supply systems like the connection to the methane network[17].

8.4.4 Profitability of actors involved

For producers of energy, apart from the investment cost, profitability is a decisive factor. Once the investment cost has been accommodated by a financial scheme, private or governmental, the operation of the project should be profitable for the

[17] This is the exemplary case of Turin where a "war between methane and DH" exists due to the absence of an integrated supply system of energy. In effect Italgas, the company engaged in the supply of methane, permits a free-of-charge connection to the methane network, while AEM (the Municipalized Energy Board) cannot do the same because of its high costs, and makes users pay 50% of the connection costs.

producer[18]. Also in this case specific observations are in order for the different technologies.

The technologies involving the use of biomass have relatively short pay-back periods (often linked to private management) and if there is the possibility of a combination with a district heating system or a connection with a pre-existent one, these technologies become more profitable for both the supply companies and for the users, who may be granted lower heat tariffs. Unfortunately this possibility seems to have been presently forgotten (the case of Milan is an example in this respect) and the most likely case to happen is a system of incineration with recover of electric energy only.

District heating technologies on the contrary, may enjoy great economic benefits due to the high efficiency of the processes, and a strong support from public administration willing to accept very long pay-back periods, typical of these technologies. As already noted, the profitability for the users could be high too because of the possibility of access to lower heat tariffs.

Of course all the previous considerations about the difficulties for possible new projects of district heating with regard to the economic and financial problems remain valid. Here it may be added that, in this case, the achievement of a large number of large-scale users exploiting the possible scale economies could also be a positive factor for suppliers. But it could be also a problem if, due to their major contractual power, they required heat tariffs lower than those ones planned for the ordinary users[19].

Furthermore, as already clarified above, under other points of view, the profitability of PV and aeolian systems is very low due to the high initial cost that cannot be repaid by the outcomes which are not efficient in the present conditions (a pay-back period of 60/80 years has been calculated for the demonstrative experiments of photovoltaic in the city of Turin). This is valid for the suppliers but also for the users if they are self-producers (otherwise if they are connected to the electric network, nothing changes).

Of course, other costs which affect the profitability of the investments for the energy producer can be included in the economic costs in addition to the technical

---

[18] In Greece most of the large projects have considerable profitability for the producer. Therefore, new initiatives are anticipated after the changes in the legal framework which support the profitability of renewable energy production. This may be proven to be an important element for future developments since it promotes also the establishment of new financial schemes based on private investors.

[19] This is in effect perceived as a really serious problem by the energy authorities of Turin.

ones already mentioned in Section 8.2. The *maintenance costs*, for example, represent an interesting variable because these are very low in the case of PV and aeolian energies, due to their 'technological simplicity' and in the case of district heating systems due to the possibility of centralizing the initiatives of maintenance in a single plant. This does not happen however, for the initiatives regarding refuse where the particularity of this kind of energy source (polluting and sometimes dangerous) imply the necessity of more frequent (thus expensive) maintenance activities.

The *location costs* on the contrary are made up by the initial costs to find a suitable place for the plants when the introduction of a new technology is decided on and to carry out studies and environmental impact analysis. They can emerge for every kind of investigated technologies, because the implementation of these technologies has always an impact on the territory, so that there may always be a problem of space or even pollution and noise factors. It is necessary to give these costs the right weight. If they are underestimated, location costs can create serious problems in the future with the risk of interrupting important initiatives only because their location had been chosen without taking into account all the consequences of this choice.

The profitability of the technology for the end-user calls for a further remark. The user perceives the economic profitability by taking into account also the convenience of the new technology. For instance, the adoption of passive solar systems for water heating would not be a profitable system in comparison to the electricity system in strict terms. The user however, perceives the advantage of having warm water all the time and under this assumption the solar system is profitable to him in economic terms. In some cases the profitability is estimated by considering the flexibility for producing renewable energy in cases where the provision of electricity is expensive, for instance, small agricultural activities, small plants, countryside and small island houses[20].

As a result, the issue of user profitability may be a very important factor. However, the user estimates his profitability by taking into account more general conditions under which renewable energy systems may show advantages in comparison with the standard system. In this respect, the losses of incomplete depreciation of the old technologies do not form a prohibitive factor.

---

[20] This issue is particularly valid in a country like Greece.

8.4.5 Conclusions on financial aspects

As already mentioned for the more developed and wider existing technologies, great efforts have been invested in R&D activities that now provide good results. In the future, however, further serious effort will become necessary due to the need for the extension of certain European-level initiatives (such as waste exploitation[21]). In fact, the development of large sized plants may create serious problems: for example future sunk costs on larger plants may turn out to be very high.

On the other hand, present excessive economic costs make the introduction of PV systems and large-scale aeolian systems impossible. It has to be stated however that the absence of wind is more a determinant for aeolian systems than the absence of an optimal insulation for PV systems. Economic costs in the first case have more chance of being lower in the future, however, the costs of aeolian resources strongly increase when there is little reduction of wind speed and when the electric network is more remote.

On the one hand, the financial conditions are better than in the past, because forms of finance exist today that were not present in the past, and these have allowed for certain initiatives that otherwise would not have been carried possible[22]. On the other hand, such forms of finance have been introduced to facilitate situations of general emergency and are often not difficult to obtain. The important question now is what will happen in the near future to those initiatives that are already in operation or have been introduced only recently thanks to temporary financial subsidies, like the ones included in the Italian CIP resolution 6/92, when these facilities come to an end. For many of these technologies in effect this resolution seems presently to be essential and the facility of 100 Lit/KWh remaining beyond eight years is not perceived as sufficient. The creation of other facilities or moreover the possibility that those that already exist become more available for all interested parties, becomes fundamental for the successful evolution of all the initiatives analysed.

Of course, PV and aeolian energy technologies have to be compared to other energy sources. It seems likely that presently - but probably also in the future - given the current expectations, alternative energy resources will have a price lower than that of non-renewable sources (see footnotes 10 and 21).

---

[21] For those technologies there will probably be strong economic costs of technical re-qualification.

[22] For example, we may refer here to the initiatives of biogas recovery from disposal, the project of PV cells in Amsterdam or the project of the district heating enlargement in Turin.

What needs to be emphasized, therefore, is the necessity to incorporate the financial participation of users, whenever possible, as well as the need for third part financing to face financial (but also technical) problems. Especially the first measure is important for technologies involving refuse and the second also for district heating implementations, at least if they are introduced recently in more optimal conditions than in the past.

Especially with regard to refuse and district heating, the future depends, paradoxically, more than in the past, on financial tools because of more serious bottlenecks and less favourable conditions.

8.5 Organizational Critical Factors

8.5.1 Political cooperation and structure of the energy supply

Political cooperation and organization of the energy supply system are very important factors determining the evolution of the initiative investigated in our analysis. It should be taken into account that renewable energy initiatives bring together several bodies and authorities for instance, local government, regional administration, electricity producers environmental authorities, private investors, users etc. These parties need to establish a new type of synergy and this is a challenging task. From the case study cities in all three countries (The Netherlands, Italy and Greece) it appears that the cooperation between these actors is crucial. All successful projects are based on a successful cooperation, while similar projects elsewhere faced a considerable delay because of lack of cooperation in this respect. What is important is the behaviour of those bodies which have legal rights to undertake and to support the initiatives[23].

A common element is the necessity of a strong political and organizational willingness to bear the high initial costs of initiatives concerning the use of refuse, the long pay-back periods and costs of the district heating technologies and especially of those technologies that at present can have only a demonstrative nature

---

[23] Therefore, it is expected that the new legal framework in Greece that promotes the decentralization and permits private initiatives will result in new initiatives under new cooperation schemes.

due to their general low efficiency. This is not a condition always fulfilled by all the investigated cities[24]; in some cases different situations exist[25].

Furthermore, there are other cases in which it is clear that the challenge has just begun. When the use of refuse as energy source is seriously taken into account, the cooperation amongst bodies most interested in the projects becomes fundamental. The bodies that have to control the observance of safety parameters and environmental standards, the bodies that deal with the waste management, and the energy companies necessarily have to cooperate.

It is possible, however, that the excessive restructuring of cooperation amongst the organizations involved, creates delays and bureaucratic problems[26].

The essential factor remains, moreover, the way in which the energy supply in the city is organized. If the supply is integrated, i.e. is in the hand of the same body, harmful forms of competition between different and alternative energy sources can be avoided. The typical example is the case of a more or less peaceful coexistence between methane and district heating, considering also their different features that makes them useful for different city areas[27].

---

[24] Probably a city which quite closely reflects this optimal situation is the Italian city of Brescia where the company that supplies methane also manages the DH system, the waste system and a lot of other services with undeniable synergies of management and technical competence. Recently, in Brescia an analysis has been carried out in order to understand which management was better for the services supplied to the city. It has shown that an integrated management can bring about an energy saving of 20% more than a disaggregated management.

[25] In the city of Pavia for example, political inactivity and bureaucracy have been for years the only reason for the delay in the start-up of an optimal district heating project to be extended in an area of interesting users. Such a project could have helped the city to face the problems stemming from the very large demand of energy (considering the role of the city as pole of a very great flow of commuters) satisfied with a not very efficient system of energy supply (gas oil is prevalent).

[26] This is the case of Turin and its initiative of biogas recovering from disposals. A high level of organizational cooperation exists between the bodies involved: AMIAT (which manages refuse) produces and sells energy to ENEL, the national energy company. AEM carries the energy to ENEL, which sells it on the market. In the opinion of AEM, however, it would have been better if AMIAT had sold the electric energy directly to them to avoid inconsistent decisions and delays, according to the national law, however, this is impossible.

[27] See the already mentioned 'war between district heating and methane' in Turin (footnote 10) and their perfect coexistence in Brescia where they are managed by the same energy company (ASM). ASM *inter alia* burns only a small quantity of methane in the district heating schemes (different from most Italian cities), as it prefers other sources and an efficient system of smoke depuration and supplying methane in the areas of the city that are less suitable to be connected to the district heating network.

8.5.2 Market conditions and timing of adoption

The energy market in most countries is - or was until recently - dominated by the national (semi)governmental public energy corporations. Such organizations were the crucial actors in any development of new activities. However, energy pricing policies did not permit the development of private activities in the sector. At the same time the inherent inflexibility of such large public bodies was a serious impediment for developing renewable energy initiatives.

New legal frameworks in the three countries investigated offer a more flexible basis, since they permit the production of energy at profitable prices for the producer. Moreover, the production from renewable resources becomes more and more an entry-free activity in which different actors can participate: for instance, local government, private enterprises, public firms and bodies etc. In this respect, it is expected that such deregulation measures promote new initiatives at least in those sectors where the conditions are mature.

In the past many successful projects were the result of actions of public bodies and these initiatives faced problems of flexibility and reliability at the decision-making level. We may assert that these problems were inherent in such initiatives which had a pioneering and very risky role. In the end, the early projects functioned as demonstration pilot projects and they prepared the market and society for more advanced steps now expected.

The market conditions can be considered contradictory and change depending on the different initiatives investigated. The market of electric energy has recently faced some problems due to restrictions in terms of the definition of the competent subjects and restrictions on the selling of electric energy produced, which can obstruct future attempts of enlargement of the initiatives[28]. Furthermore, in the case of district heating, high engineering costs and more competitive methane make the adoption processes more difficult in general, while on the other hand, the market of operators for refuse-related technology is not very big, considering the high technical specialization required.

However, if there is one case of a really problematic situation, then it is the case of PV systems. This market is absolutely not free: only a few actors spend efforts in R&D, only a few actors produce and install the technologies and the prices of alternative energy sources are (still) very competitive with prices of PV energy.

Cases of optimal exploitation of *external market situations* have taken place in the past with regard to cases of implementation of district heating systems. This is for

[28] Storage problems would arise in this case.

example the case with the Italian city of Brescia and the Dutch city of Almere, which in the seventies were able to take advantage of the high prices of traditional energy sources and the high inflation lowering the loan value, by starting up one of the earliest district heating plants in these countries.

8.5.3 Management and technical learning processes

Management and learning costs are present, even if they are not considered to really be factors which delay the adoption process. They are minor in technologies that are simple at present, like solar water heating. For the other technologies, such costs are higher when initiatives are still in an experimental phase, while they emerge especially at the beginning of projects. This consideration may not be valid for district heating applications, even if it is true that to switch to 'European sizes' now means facing high costs of management and expensive learning processes. With technologies that make use of energy sources like refuse, these costs are directly linked to the elements analysed above, i.e. the organization of the energy supply in the city, the separation or integration inside the city of the competences about the different energy sources, and the possibility of resorting to tools such as part financing to give external impetus to the management of initiatives.

8.5.4 Conclusions on organizational aspects

The evolution of organizational processes is very important for the future success of most initiatives analysed in our field study. It will be necessary to take further into account the *bargaining costs* between all the actors involved in those initiatives (not only the public bodies or the private companies but also the *passive* actor, the population), in order to avoid the negative consequences of a lack of a cooperative culture. Those costs, if born, can represent a long term success factor.

Furthermore, it is clear that cooperation and exchange of competences will be more necessary, especially in those cases where an integrated energy system does not exist and different public bodies are entrusted with the task of offering energy from different sources. This will help face and overcome the management costs and the learning costs stemming from the start-up or the enlargement of new initiatives.

The longer cities wait with the introduction of innovative technologies for energy production, the more difficult the situation becomes. This is especially so as far as the use of refuse as an energy resource and its connection to a district heating system (the latter more improbable than the former), are concerned. Here the way in which the overall system of waste management is organized, is important; other mistakes must also become unacceptable. The mistake made by some cities in not starting with this technology when the waste emergency emerged for the first time, has already caused high sacrifices in terms of money and general discomforts[29].

8.6 Ecological and Social Critical Factors

8.6.1 General and specific environmental awareness of population

Over the past years we have observed an increase of a general environmental awareness of the population which often turned against those initiatives with more visible impact. But it has also exhibited more possibilities for cases where the great effort of competent actors in reducing negative impacts, or moreover illustrating positive aspects (energy savings and related economic savings for users)[30], has been appreciated. For future initiatives, environmental awareness may play an even more important role, once the economic conditions will no longer be in favour of the non renewable energy technologies.

In fact, the involvement of the *passive* actors, and also the recognition of their role as *active* in the decisional process for the start-up of a new initiatives, has always shown positive effects and demonstrated its necessary presence in helping avoid that the population delays certain activities, which make it impossible for them to continue.

---

[29] To understand the seriousness of the situation, it is sufficient to consider the case of Milan where refuse has not only always been sold off in disposal sites in the own area, but also has been transported to disposal sites of other cities. This fact of course has caused an unsustainable situation with a large discomfort for the population and high costs due to the outside transport of refuse and the financial compensations provided for other cities bearing these additional refuses.

[30] This statement probably needs some nuances for different countries. For example, it seems that no large initiative in Greece has been based on the environmental awareness of the population. Here other reasons, mainly economic and convenience, played a crucial role. Environmental awareness was a supporting element for an information campaign in favour of the new technology, however it was not the decisive factor. In contrast, we mention some renewable energy projects in The Netherlands (e.g. the solar cells in Amsterdam and solar water heating in Gouda) which rely to a large extent on the environmental awareness of the participants.

Furthermore, the non-opposition to other potential technologies (which is the case of aeolian and PV energies), thanks to the absence of pollutant impacts, turns into an indifference in the choice of a possible adoption of these technologies, due also to the lack of marketing and communication activities by the entrusted companies.

8.6.2 Social benefits of technologies

There are many advantages related to the investigated renewable energy technologies, but it is interesting to make a distinction between the *objective* advantages (supply side) and the advantages as they are *perceived* by the population (demand side). The objective advantages are in general linked to the savings of energy, the use of alternative energy sources which are often (but not always) cleaner, the reliability of the technologies, and sometimes the reduction of air pollution. The public however, does not attach the same importance to these advantages and perceives them in another way. Economic considerations have a great value for users. Therefore, the public can remain collectively rather passive in the light of environmental and energetic advantages, if these advantages are not accompanied by economic savings (like the costs of energy or the costs of connection).

The initiatives that meet most hostility by the public remain anyway those technologies linked to the management of refuse and especially when they are located in operation nearby populated areas. It does not matter whether the incinerator plant actually reduces the amount of refuse going to the disposal, or if the plant is equipped with installations that reduce the negative effects of an uncontrolled leakage of biogas.

This depends on the diversity of the investigated initiatives in terms of environmental impacts. All these initiatives have (or could have) a great impact, but sometimes this is only an impact on the territory or in terms of noise, when the technologies are essentially clean. Sometimes, however, there are real impacts on the quality of the air, when an energy source is burnt (gas oil, methane or, worse, waste)[31].

This explains why it becomes important to inform the population as best as possible of all advantages of these kinds of initiatives. Moreover, it is necessary that the management is serious and accurate, and chooses the locations accurately.

A common discourse also regards the harmony with the territory or with pre-existent structures. Harmony with the local living area depends on the size of the plants, the care with which locations are chosen, and the impact on local citizens. Sometimes the location of some initiatives is unavoidable[32]. Furthermore, it depends sometimes on the different points of view. The district heating is once again an example: a harmony exists for houses that are connected to the district heating network, but it is clear that the plants where the energy sources are burnt, especially when they are large, have a strong impact on houses in the direct surroundings.

For most of the successful projects in The Netherlands, Italy and Greece the new technology was fully compatible with the structure of the existing buildings[33]. Therefore, this factor did not create an obstacle in any respect. However, in the future, this issue may become more crucial for the adoption of PV systems and insulation systems in existing buildings. It seems that a legislative basis is required for imposing certain standards for the implementation of these technologies especially in the existing urban structures. The existence of a clear legal basis would promote also the adoption of these kind of initiatives.

---

[31] The example of district heating can be useful to understand this consideration. District heating has great advantages, because it is easier, efficient and less expensive to control one big chimney for the waste smoke rather than a lot of little chimneys badly managed. It is also true however, that all this depends on the source burnt in the system. If we burn methane we are sure to burn the most clean non-renewable energy source but also the most precious one. If, on the contrary, we burn fuel oil or even coal we burn more economical sources but also the most polluting ones, as then we have to plan an efficient ex-post depuration system. This is especially valid if we burn refuses: in this case we use a source available almost without costs but, considering its variability and instability, costly processes of 'inertization' and ex-ante treatment of refuses and processes of ex-post depuration become essential.

[32] This is the case of biogas recovery from disposal where the location of engines and plants depends on the previous choice of simple disposal, and whether the problems stemming from a disposal close to inhabited areas, was taken into account. These problems remain regardless of the positive effects in terms of pollution and safety related to the recovery of biogas.

[33] For example, in the PV case of Amsterdam, the solar cells have been integrated in the building structures in an unusual, but very attractive way.

8.6.3 Conclusions on ecological/social aspects

The difference between relatively clean technologies and those technologies that, while being efficient, are polluting, has to be underlined, because in the future we will have necessarily to resort more and more to those technologies involving the use of refuse. At present, these are the most problematic technologies in environmental terms. It is for this reason then that the necessity emerges to devote more attention in the future to actors who seemingly are passive but in real fact are most interested in the start-up of those initiatives implying energy saving and pollution improvements.

Another problem may emerge from the (lack of) available space. The increasingly evident lack of available and suitable space for renewable energy installations forces policy-makers to give more attention to research on more efficient processes to supply a greater quantity of energy with the same size plant. This consideration is even more important for PV energy systems and especially aeolian systems, because at present these plants require a very large size to be efficient in terms of energy saving, a size that however is not only unsustainable in terms of impacts on the territory but also in terms of economic costs.

9 A Comparative Empirical Analysis of Urban Energy Policies: a Meta-Analytic Approach

9.1 Introduction

The previous chapter aimed to give a qualitative overview of success and failure factors involved in renewable energy policies in various European cities. In the present chapter we will systematically compare across the various renewable energy technologies envisaged, the experienced and expected degrees to which these aspects act as a bottleneck or facilitate factors for the introduction or success of these technologies. In addition, we aim to compare also the potentials of these technologies to contribute positively to urban sustainability.

The information source for these comparisons is provided by the same experts who have been interviewed in the qualitative analysis (see Section 9.1). These experts have also been asked to give quantitative scores for the levels to which the various technologies contribute positively to urban sustainability, and for the experienced and expected *magnitudes* of the levels to which all these aspects are regarded as impediments to the adoption of these technologies. This information has been systematically collected by means of a common checklist which has been used throughout the field work.

Each of the respondents then has given scores with regard to both existing projects and implementations of renewable energy technologies for his city, as well as technologies that have not yet been implemented but might be introduced in the future. Thus we distinguish in our analysis 'real cases' and 'virtual cases' of technologies. The analysis thus not only investigates experienced problems related to existing technologies, but also expected problems with regard to hypothetical situations for the case cities. The experienced problems may be seen as factors influencing the success rate of existing technologies, while the expected problems may be seen as factors obstructing any investments in renewable energy technologies in the cities. Together, information has been collected on sixty situations, each of which concerns a (real or virtual) combination of a particular technology in a particular city.

The chapter is organized as follows. Section 9.2 deals with the contribution of the various renewable energy technologies to urban sustainability. This is done in a descriptive way, by means of using average scores. Section 9.3 identifies the strong and the weak points regarding the distinguished critical success factors for the various technologies. This is also done in a descriptive way, again by using average scores.

This descriptive part of analysis is based on aggregating scores provided by the interviewed experts by taking their mean values across the observations. Clearly, in a strict methodological sense, such scoring data are not entirely suitable for being aggregated in this way. A technique like multidimensional scaling might be used in such cases. However, the aim of this chapter is to illustrate and highlight differences in scores between different potential bottleneck factors and between different technologies. For this aim, the used method may be justified.

In Section 9.4, we present a second line of our quantitative analysis, in which we identify the main factors that are responsible for the success of the initiatives. For all technologies together, we have sixty cases, which is a sufficient number to apply standard statistical techniques. In our case, we will employ an ordered probit model, as most of our data are ordinal in nature. In order also to have results for each technology separately, we have to resort to another kind of analysis, as the number of observations per technology is not sufficient for a standard statistical analysis. In this context, *meta-analytic techniques* may offer an interesting analytical contribution towards a better understanding of the critical success factors of urban sustainability policies (see Van den Bergh *et al.* 1997). Meta-analysis serves to synthesize research findings from previous studies or to formulate transferable policy conclusions from previous policy experiences. The meta-analysis presented in this chapter has therefore an exploratory character. We will make use of expert opinions on the performance of various energy and environmental technologies in several European cities. This forms a contrast with conventional meta-analytic studies that make use of different empirically-based impact studies on the same phenomenon. Nevertheless, the results may be of great interest for sustainable city initiatives.

9.2 Positive Impacts on Urban Sustainability

9.2.1 All technologies

In our analysis we will use three indicators for urban sustainability related to renewable energy technologies. The first indicator is the production of energy by these technology applications. This in fact is equal to the saving of depletable energy resources. The second indicator is the reduction of pollution caused by (conventional) energy production. Both indicators are relevant proxies for environmental sustainability. The third indicator is an approximation for economic sustainability. It relates the production of energy by renewable energy systems to the necessary economic resources.

The scale which has been used for collecting information from the respondents on the magnitude of these indicators is a five-point scale. The lowest possible score is 1, referring to respectively a low energy production, a small pollution reduction and a low economic efficiency ratio. The highest possible score is 5, referring to a high energy production, a large pollution reduction and a high economic efficiency ratio.

In Figure 9.1 we have shown the average scores of the experiences and expectations regarding the sustainability aspects of renewable energy technologies. The general feeling from these findings is rather optimistic as the mean scores are above the middle value. The area in which relatively most benefits are to be obtained is generally perceived to be pollution reduction, while the production of energy comes second. This may be explained by the fact that some technologies are still in an infant phase and can generate only limited amounts of energy (PV, CWE).

The size contrast between the indicator scores is larger for the virtual cases. Especially the expected reduction in pollution appears to be higher than the experienced reduction.

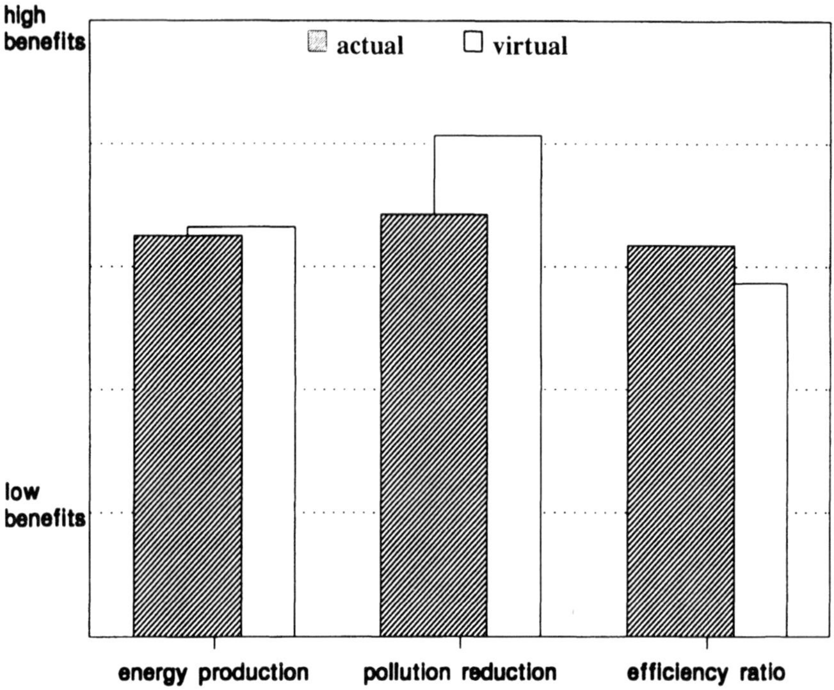

Figure 9.1. Average scores of sustainability indicators for existing real cases and hypothetical (virtual) cases ('low benefits' refer to a score value of 1, 'high benefits' refer to a score value of 5)

9.2.2 Impacts by each technology

In Figure 9.2 the experienced and expected scores for sustainability aspects for each renewable energy technology are shown. Some interesting conclusions can be drawn. First, it appears that those technologies which really utilize renewable sources (PV, SWH, CWE, Biogas, DH/R) have relatively low pilot results concerning their capacity to generate energy. An exception is refuse incineration, which shows a higher value together with CHP/DH (using fossil sources) and energy saving technologies. It can be concluded that the technologies that have been mostly developed until now, simply show the highest energy production potential. On the contrary, it is the technologies that have been least developed until now (PV, SWH, CWE), that show the highest benefits in the area of reduction of pollution.

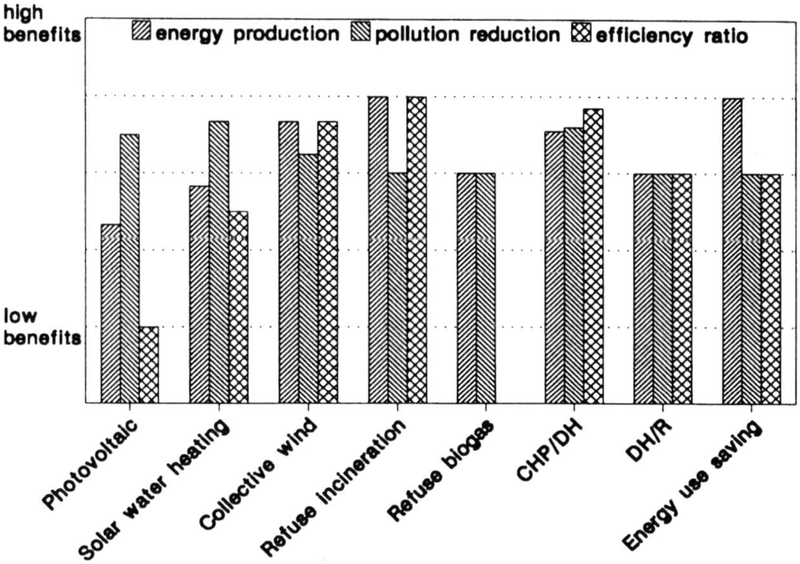

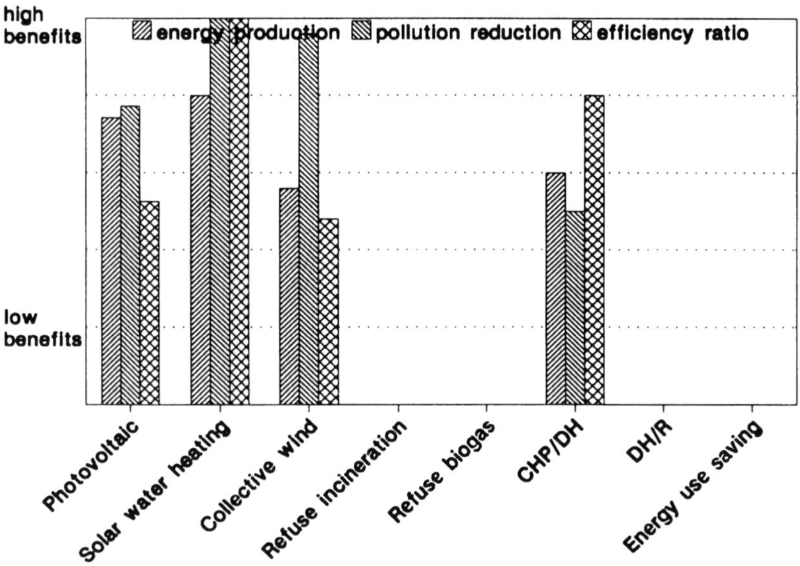

Figure 9.2. Average scores of sustainability indicators by renewable energy option, for existing cases and virtual cases ('low benefits' refer to a score value of 1, 'high benefits' refer to a score value of 5)

The experiences with refuse biogas and district heating using renewable sources are not so favourable in relation to the other technologies. Both these technologies have only intermediate scores for energy production and pollution reduction.

It is interesting also to look at the economic efficiency of the various existing applications. This pattern clearly explains the frequency in which technologies are actually being implemented. Most efficicient are refuse incineration and CHP/DH. Both are applied in many cities on a large scale and make up a significant share of these cities' energy supply. Moreover, there are several technologies with only intermediate scores on economic efficiency (SWH, CWE, DH/R and energy use savings). Of these latter, SWH and energy use savings are the technologies applied on relatively large scales. PV technologies show the smallest economic efficiency which is mainly due to its actually low energy generation capacity.

When we compare these real experiences with expectations in the cases of virtual implementation, we discover a difference between solar applications on one hand and wind and CHP/DH systems on the other. In the first case, expectations of all three sustainability indicators are higher than the real experiences, while this is not the case for wind and CHP/DH systems.

9.3 Emerging Bottlenecks

9.3.1 All technologies

In this section we systematically compare for the whole set of renewable energy technologies the experienced and expected *magnitudes* of the levels to which all these aspects act as bottleneck factors to the introduction or the successfulness of these technologies.

The information source for these comparisons is provided by experts who have been interviewed in the case cities of the project. These experts have assigned quantitative scores to the experienced (real-world applications) and expected (virtual applications) attitudes of the levels to which the relevant aspects are seen as bottleneck factors to the introduction or the successfulness of these technologies.

For the scores on the critical key aspects judged by the various local experts, the following rank order scale is used:

5: the aspect concerned is not a barrier at all; on the contrary, the way in which the condition is met is regarded as a reason for the success of the application of the technology;
4: the aspect is a condition that is sufficiently met by the application of the technology;
3: the aspect is neither a clear positive nor a clear negative critical factor, nor it is irrelevant;
2: the aspect is a barrier which with some effort may be overcome in the near future;
1: the aspect acts as a major bottleneck.

In Figure 9.3 the average scores of the critical success factors for each group of Pentagon factors are shown. It appears that - in a relative sense - the main bottleneck factor is the financing of the implementation of technologies. This result is in harmony with the proven influence of the economic efficiency ratio, as seen also in the previous section. Note however that even this largest experienced bottleneck factor still has an intermediate score (not really an obstacle, neither a condition that is sufficiently being fulfilled by the city). The score is however significantly lower in the expected case.

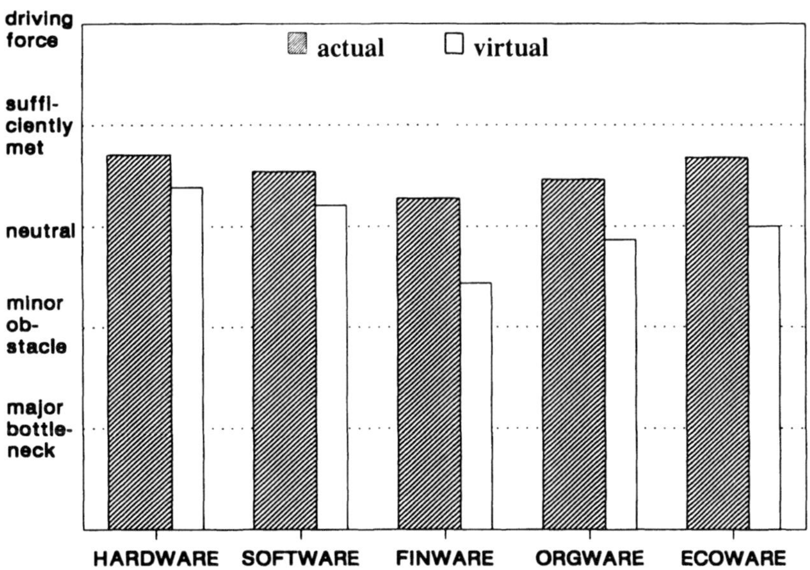

Figure 9.3. Mean scores for the five groups of Pentagon factors, experienced and expected

The picture shows that the field of organizational structure regarding the implementation of technologies takes second place in main bottleneck factors. Requirements related to the technology itself constitute in general the least problems. The contrast between these mean scores of organizational structure and the technology itself may confirm that the implementation process of renewable energies is merely a technology-driven process.

Furthermore, it is interesting to see that in general the scores of experienced bottleneck factors offer a more optimistic picture than in the expected cases. This is valid for all distinguished factors.

The comparison of mean scores between hardware factors lead to some interesting conclusions (see Figure 9.4). It appears, for instance, that the climatic conditions constitute one of the smaller problem areas. This is a logical finding, as renewable energies typically require specific climatic conditions in order to operate efficiently, and cities choose for the most suitable technology for their case. In addition, the purely technological state generally also shows few problems. This regards both the actual technological know-how of renewable energy technologies in general as well as the technological compatibility with other (existing) applications. It is interesting to see that in the expected case the size contrast between the scores of the technological state and other factors is even larger.

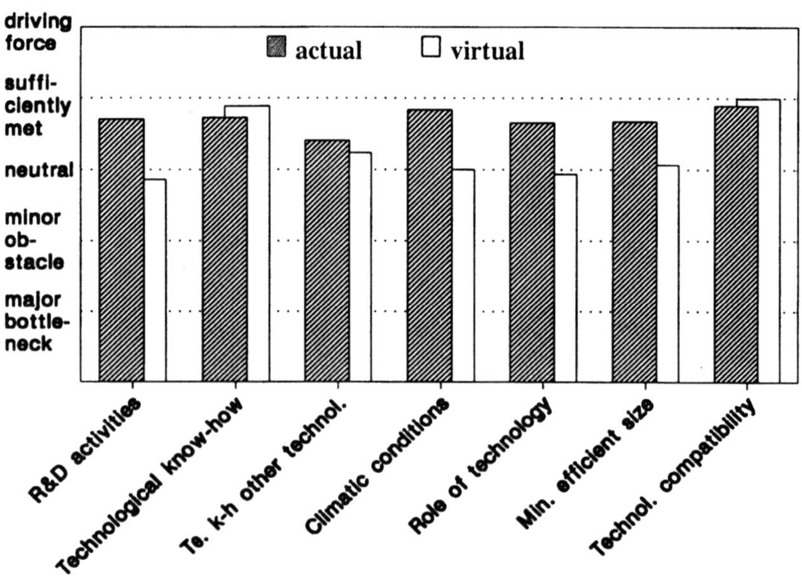

Figure 9.4. Average scores of hardware critical success factors

In contrast to this, the technological know-how of the renewable energy linked technologies constitutes a relatively large problem. This issue may be illustrated by the PV case, in which the successful installation of PV systems on buildings not only depends on the technological features of the PV systems themselves, but also on the quality of other necessary building components and the know-how of the constructors. In some cases the role of these linked producers and constructors has been underestimated.

In Figure 9.5 mean scores of the software critical success factors are shown. They illustrate a larger variation than the hardware factors. There is a general experience that the reliability requirement of renewable energy systems for the end-user is sufficiently met in existing applications. In addition to this reliability, there are quite positive scores for the adaptation by households and large users to the use of these systems. Other factors related to the user however, show only intermediate scores. Although generally these factors cannot be marked as (minor) obstacles, there is a general feeling that the residential awareness of renewable energy systems should improve. This limited awareness has its reflection also in the reaching of a substantial (critical) number of end-users.

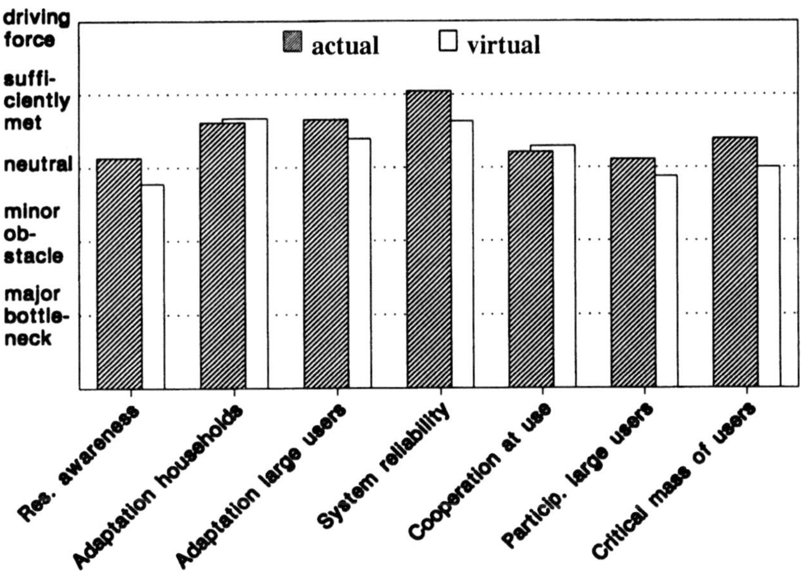

Figure 9.5. Average scores of software critical success factors

Scores for financial apsects are depicted in Figure 9.6. The scores in the experienced case show large similarity across the various aspects, with a lowest score for the sunk costs aspect. The financial aspects show, to even a larger extent than technological, user-related and organizational aspects, a difference between actual experiences and expectations, the latter being more pessimistic. In particular, there is a general ex-ante fear that investment resources in renewable energy technologies do not pay back themselves. This aspect has clearly the lowest score in the non-case.

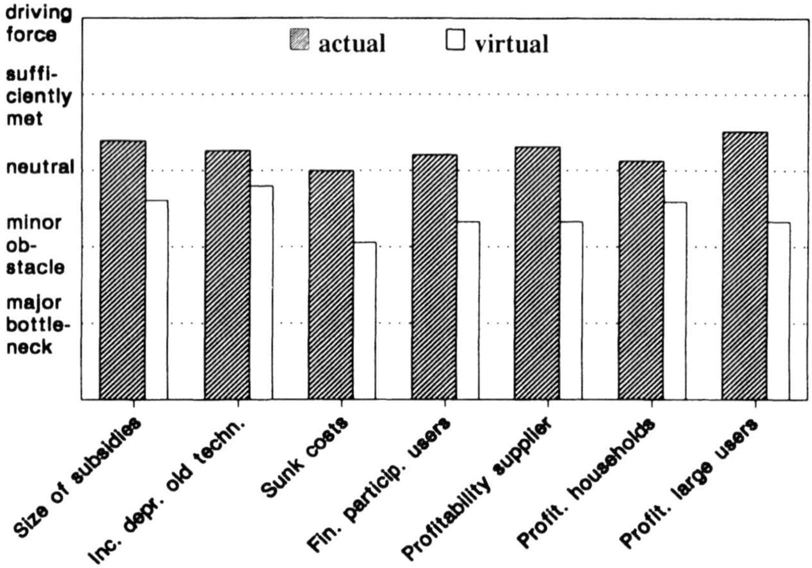

Figure 9.6. Average scores of finware critical success factors

Concerning organizational aspects, the (external) market conditions in which initiative takers have to operate show relatively low scores in comparison with other aspects (Figure 9.7). An underlying relevant issue here is found in the on-going liberalization of the energy supply market in the countries investigated, which may be seen as a major reason for the restrictive character of market conditions on renewable energy projects.

Generally, the score variation between organizational aspects is not large in the cases of actual implementation. The variation is larger for the expected (virtual) case. In particular, the expectations show that in addition to a low score for market conditions, requalification of existing staff is perceived as a minor obstacle. Actual experience shows a more optimistic view on this aspect.

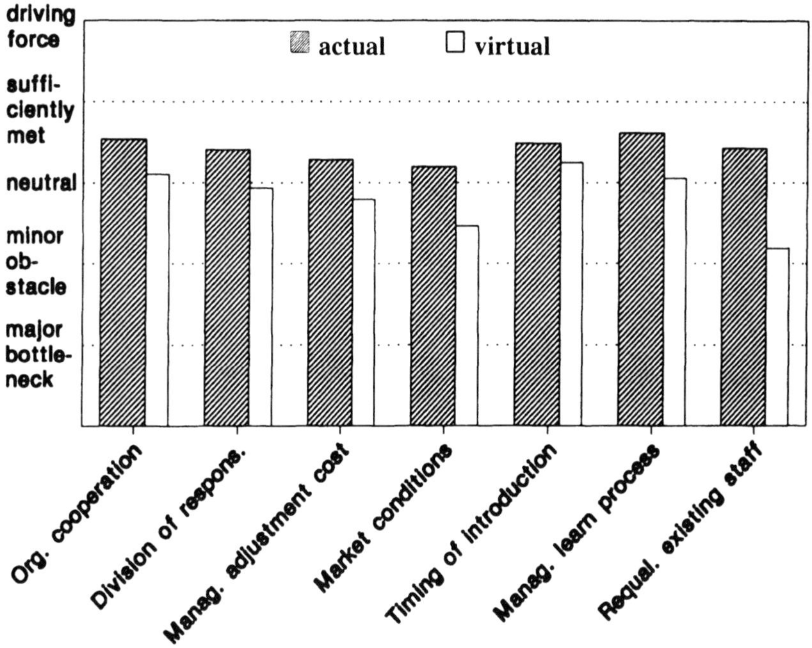

Figure 9.7. Average scores of orgware critical success factors

In Figure 9.8 the scores of ecological side-aspects of renewable energy policies are depicted. It appears that a large potential is assigned to the introduction of renewable energy technologies to yield also so-called non-tangible benefits. In contrast, environmental awareness among the public scores relatively low. In general, we may state that environmental awareness and awareness of renewable energy systems are only moderately fulfilled conditions.

The picture of ecoware factors shows, just as in the case of financial aspects, large differences between the experienced and expected case.

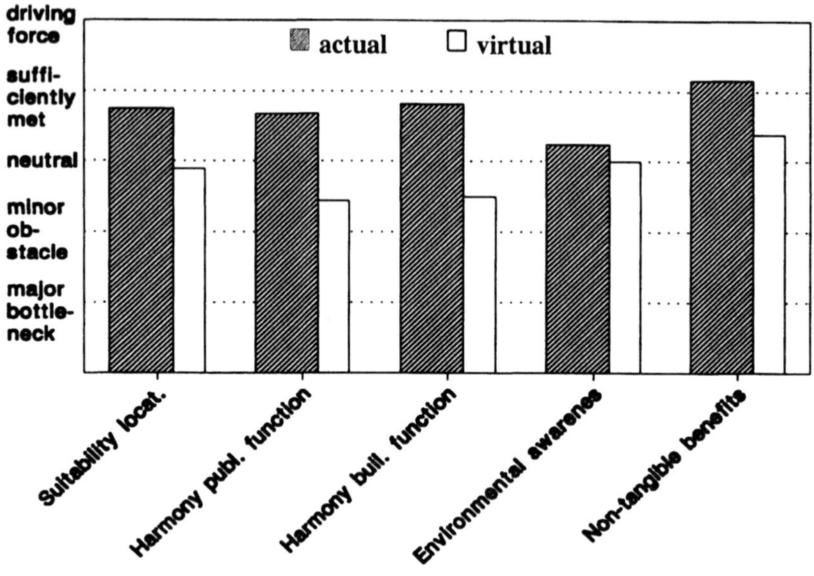

Figure 9.8. Average scores of ecoware critical success factors

9.3.2 Emerging bottlenecks by each technology

After the general conclusions in the previous section in which all technologies were taken together, we will now focus on the differences between the various technologies concerning the five groups of Pentagon factors (see Figure 9.9).

Looking at the technological aspects, we see a difference between on the one hand PV, SWH and CWE systems, and on the other hand refuse incineration and biogas, CHP/DH and DH/R. It is experienced that technological factors constitute relatively large problems for the first group in comparison with the second group. This is not hard to understand, facing the still immature technologies of PV and CWE.

District heating systems appear to score very well on user-related aspects. Reasons may be found in the fact that district heating requires no adaptation efforts of the end-user, that it has a high degree of comfort/reliability, and that a relatively high number of users can be reached by district heating infrastructure. Relatively lower scores are for the PV, SWH, CWE and energy use saving technologies.

The mean scores of the financial aspects in the experienced case can be explained as follows. The highest scores are for energy technologies that make use of refuse and district heating applications. These technologies show already a relatively large economic feasibility. The other technologies have scores which are significantly below the intermediate value. For SWH and energy use saving technologies, this could be explained by the dependence on private financing; for PV and CWE technologies, this could be explained by their low efficiency rates.

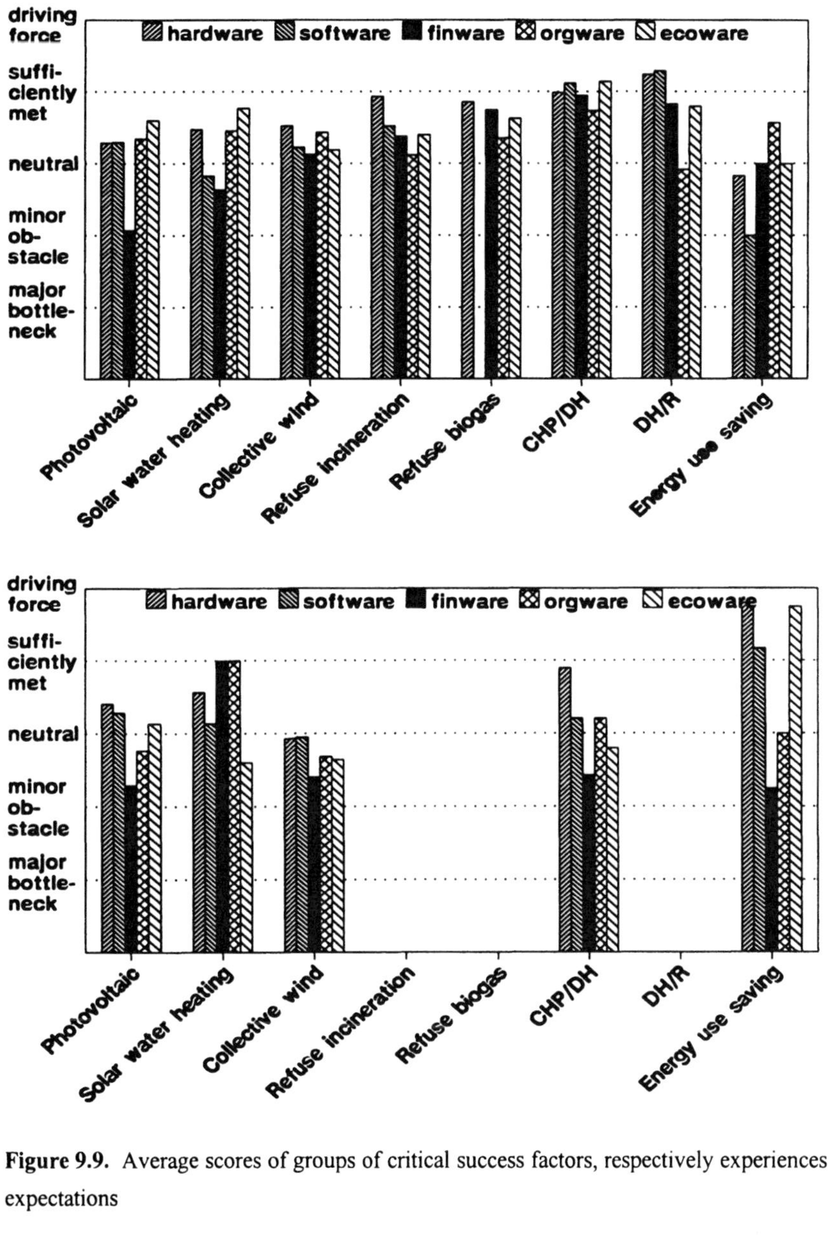

Figure 9.9. Average scores of groups of critical success factors, respectively experiences and expectations

In the area of organizational structure, the introduction of two systems score relatively well compared with the others, namely the introduction of SWH and CHP/DH systems. In both of these cases there is already much implementation experience, and it seems that this has a beneficial impact on the quality of the organizational structure probably due to a relatively long learning process.

Finally, we may notice that collective wind energy shows - together with energy use saving - a relatively low score on ecoware. Such a finding can be easily explained by the fact that wind mill parks in general meet much public resistance because of their negative impacts on the local environment: landscape deterioration, harm to specific public functions like natural destinations etc.

When we compare the experienced case with the expected case, we see that in particular there are two renewable energy technologies that show more positive values in the experienced case, namely solar water heating and CHP/DH. The inverse is true for energy use saving, which in practice proves to face higher bottlenecks than expected.

9.4 Investigation of Key Factors Influencing the Success Rate

In this section we seek for explanatory factors for the level to which new urban energy technologies are successful. The (perceived) rate of success is assumed to depend on how the technologies score with regard to the critical factors from the Pentagon model which have been distinguished in the former. For instance, it seems plausible that the probability that an already implemented technology in a city will be expanded, is larger when there exist no severe bottlenecks on the financing of the expansion of these investments.

The level of success of a certain technology in a given city is measured by inquiring respondents on the probability that an existing technology will be implemented at a larger scale in their city, or that new investments will be made in an as yet not implemented energy technology. The perceived success rates are assessed on a categorical scale from 1 to 5 (from a very low to a very high probability to enlarge the technology implementation or to start new investments in it).

As methods we use standard statistical analysis as well as a newly developed method called meta-analysis. The latter is necessary for the analysis of the perceived rate of success at the level of a separate technology, as the number of observations is in this case not sufficient for standard statistical analysis.

9.4.1 All technologies: a probit analysis

The data base on the sustainable city initiatives in our sample is rich. From a methodological perspective, it is interesting to test the relevance of the five key factors incorporated in our Pentagon prism. The number of observations (60) allows a statistical regression analysis with the (perceived) rate of success of the technologies as a dependent variable, and the critical Pentagon factors as independent variables. As all our data are of an ordinal nature, we will utilize the ordered probit model in our applied analysis.

The analysis will be carried out in two steps, viz. at the overall level of the Pentagon factors and at the level of individual components of the Pentagon factors. First, we will carry out the analysis for the main groups of Pentagon factors. In this case, we have five independent factors. Each of these is quantified by taking the average score of the critical factors belonging to the respective main group of Pentagon factors (where we implicitly assume that the intervals between the score values 1 to 5 are equal). We have chosen for this pragmatic approach of aggregating the values of the elements of these factor groups in order to obtain a tentative insight into the relative influence of the main Pentagon factors (for instance, the influence of financial factors compared to hardware factors). Furthermore, in our analysis we use the original measurement scale of the dependent variable, viz. the perceived success rate, which is a categorical (ordered) scale ranging from 1 to 5.

In Table 9.1, the results of the ordered probit analysis are shown which compares the influences of the five groups of Pentagon factors on the perceived success rate of renewable energy technologies. The majority of the five factors appears to have a positive coefficient. This positive influence was to be expected for the reason that the less stringent the bottleneck factors are, the higher the probability of a successful implementation will be. Two positive coefficients appear to be significant: these concern hardware and finware. This tentative investigation thus indicates that - in addition to the technological impact of renewable energy options - financial aspects appear to be, in a general sense, decisive for the potential success of the renewable energy technologies in cities.

In a second stage, we have undertaken a more rigorous analysis by carrying out the analysis for each of the constituents of the main groups of Pentagon factors. For each of these groups we have run a separate ordered probit analysis. In these cases, we can take full account of the ordinal character of the data by creating dummy variables for relevant categories of the score values. For this purpose, we have created three classes of score values. The reference class is the category of score values 1 and 2 (reflecting respectively major bottlenecks and modest obstacles). Two dummy variables (D1 and D2) are next created, reflecting respectively that the aspect is neutral/not relevant and that the aspect is sufficiently met or even a driving force for the implementation of the technology at hand. Table 3.1 in Chapter 3 will be used as the overall framework for our test.

Table 9.1. Probit results for sustainability success rates with main Pentagon groups as independent variables

Variable (means)	Probit Estimate	Std. Error	t-value	p>|t|
Hardware	0.47524	0.2661	1.79	0.080
Software	-0.44192	0.2459	-1.80	0.078
Finware	0.41688	0.2338	1.78	0.081
Orgware	0.26103	0.2788	0.94	0.354
Ecoware	0.00865	0.1955	0.04	0.965
Constant				
Alpha_1	0.78893	0.8861	0.89	0.377
Alpha_2	1.23062	0.8963	1.37	0.176
Alpha_3	2.13628	0.9540	2.24	0.030
Alpha_4	3.03258	0.9882	3.07	0.003
Measures of fit:				
Likelihood Ratio Chi-square with 5 d.f., prob= 0.0113:				14.7923
-2 Log Likelihood for full model:				162.0846
-2 Log likelihood for restricted model:				176.8769
Percent Correctly Predicted:				36.6667

We will now concisely evaluate the results of our ordered probit analysis, as given in Tables 9.2-6. When we focus only on the *hardware* factors in order to explain the success of renewable energy systems, it appears that not much can be said on the individual influence of each of these factors when they are compared with one another (see Table 9.2).

Table 9.2. Probit results for hardware factors

Variable		Probit Estimate	Std Error	t-value	p>|t|
R&D activities	D1	-0.39272	0.7765	-0.51	0.616
	D2	-0.51650	0.7974	-0.65	0.521
Technological know-how	D1	0.14737	1.2155	0.12	0.904
	D2	0.84383	0.9964	0.85	0.402
Technical know-how of other	D1	0.23373	0.7743	0.30	0.764
technologies	D2	0.80192	0.7385	1.09	0.284
Climatic conditions	D1	0.91308	1.0024	0.91	0.368
	D2	1.15468	0.9317	1.24	0.222
Role of technology	D1	0.47341	0.5261	0.90	0.373
	D2	0.43474	0.6331	0.69	0.496
Minimum efficient dimension	D1	-0.48104	0.5816	-0.83	0.413
Compatibility with existing	D2	-0.74503	0.5632	-1.32	0.193
technologies	D1	-0.26638	0.7124	-0.37	0.710
	D2	-0.41625	0.6283	-0.66	0.511
Constant					
Alpha_1		-0.44330	0.9789	-0.45	0.653
Alpha_2		-0.04165	1.0108	-0.04	0.967
Alpha_3		0.87348	1.0368	0.84	0.404
Alpha_4		1.86949	1.0475	1.78	0.082

Measures of fit:	
Likelihood Ratio Chi-square with 14 d.f., prob= 0.4883:	13.4913
-2 Log Likelihood for full model:	160.8517
-2 Log likelihood for restricted model:	174.3430
Percent Correctly Predicted:	50.0000

Roughly half of the independent variables turn out to have a positive coefficient compared with the base. In this particular case, none of the separate factors can be qualified as really significant, but from the factors with positive coefficients compared with the base, it appears that the relatively highest significance is offered by the climatic situation of the site. This suggests that the location climatic conditions which affect the efficiency of the energy technologies in the city, have a larger beneficial impact on the success of the technologies than the technological state of these technologies alone. However, it has to be admitted that the significance of this factor is not impressive.

For the *software* factors, the results are more conform to our expectations. In this case the majority of the variables have a positive coefficient compared with the base (see Table 9.3), while these are in most cases higher for the second dummy variable. The participation of large users appears to have the largest significance, which seems to be an interesting and plausible result.

Table 9.3. Probit results for software factors

Variable		ProbitStd. Estimate	Error	t-value	p>|t|
Public awareness	D1	-0.65019	0.6652	-0.98	0.334
	D2	0.23284	0.7584	0.31	0.760
Adaptation of households	D1	0.00629	0.9030	0.01	0.994
	D2	0.31797	0.8791	0.36	0.719
Adaptation of large users	D1	0.11229	1.0584	0.11	0.916
	D2	-0.94098	1.0567	-0.89	0.378
Reliability for end-users	D1	0.63280	0.8940	0.71	0.483
	D2	0.57865	0.9305	0.62	0.537
Cooperation in use	D1	0.41753	0.6869	0.61	0.547
	D2	0.56519	0.7093	0.80	0.430
Participation of large users	D1	0.20003	0.5559	0.36	0.721
	D2	1.13786	0.6863	1.66	0.105
Critical mass of users	D1	-0.14372	0.4429	-0.32	0.747
	D2	0.15026	0.5843	0.26	0.798
Constant					
Alpha_1		-0.78051	1.3241	-0.59	0.559
Alpha_2		-0.32819	1.2211	-0.27	0.789
Alpha_3		0.63814	1.1729	0.54	0.589
Alpha_4		1.62452	1.2071	1.35	0.186
Measures of fit:					
Likelihood Ratio Chi-square with 14 d.f., prob= 0.3330:				15.6847	
-2 Log Likelihood for full model:				158.6582	
-2 Log likelihood for restricted model:				174.3430	
Percent Correctly Predicted:				41.6667	

In the case of *finware* factors, the picture is again less clear, although there is one variable with a positive and significant contribution, and this is the profitability of the suppliers (Table 9.4).

Table 9.4. Probit results for finware factors

Variable		Probit Estimate	Std. Error	t-value	p>|t|
Initial governmental subsidies	D1	0.45417	0.7741	0.59	0.561
	D2	0.38871	0.6868	0.57	0.574
Incomplete depreciation of old technologies	D1	0.28883	0.6227	0.46	0.645
	D2	-0.27169	0.9899	-0.27	0.785
Sunk costs	D1	0.00797	0.5157	0.02	0.988
	D2	0.02262	0.7041	0.03	0.975
Financial participation of users	D1	-0.26673	0.4111	-0.65	0.520
	D2	0.43133	0.4970	0.87	0.390
Profitability for supply actors	D1	0.84936	0.5065	1.68	0.101
	D2	0.78931	0.5882	1.34	0.187
Profitability for households	D1	0.36331	0.5032	0.72	0.474
	D2	0.93466	0.9308	1.00	0.321
Profitability for large users	D1	-0.51745	0.7217	-0.72	0.477
	D2	-1.03956	0.8230	-1.26	0.213
Constant					
Alpha_1		-1.23662	0.7634	-1.62	0.113
Alpha_2		-0.82362	0.6700	-1.23	0.226
Alpha_3		0.15686	0.6786	0.23	0.818
Alpha_4		1.16081	0.7081	1.64	0.109

Measures of fit:	
Likelihood Ratio Chi-square with 14 d.f., prob= 0.4074:	14.5803
-2 Log Likelihood for full model:	159.7627
-2 Log likelihood for restricted model:	174.3430
Percent Correctly Predicted:	31.6667

Looking at *organizational* factors (Table 9.5), it appears that one variable has a clearly significant (positive) contribution and this is offered by the market conditions (second dummy variable). This finding indicates the importance of external conditions created by other actors for the success of renewable energy projects, which appears to be more critical than organizational matters that are internal to local sustainability projects.

Finally, we take a look at social/environmental factors in the '*ecoware*' component of the Pentagon model (Table 9.6). Analysing these factors separately, it appears that the harmony with public functions of a specific site for the renewable energy technology has a significant, positive coefficient for the first dummy variable. Also this result seems to offer a plausible explanation.

Table 9.5. Probit results for orgware factors

Variable		Probit Estimate	Std. Error	t-value	p>|t|
Organizational cooperation of supply	D1	-0.45356	0.7391	-0.61	0.543
actors	D2	0.04093	0.7697	0.05	0.958
Division of responsibilities	D1	0.11932	0.7181	0.17	0.869
	D2	0.38687	0.7880	0.49	0.626
Adjustment costs of management	D1	0.13576	0.8202	0.17	0.869
	D2	-0.71071	0.9057	-0.78	0.437
Market conditions	D1	0.14155	0.4768	0.30	0.768
	D2	0.85581	0.4509	1.90	0.065
Timing of introduction	D1	0.77852	0.5788	1.35	0.186
	D2	0.27013	0.4810	0.56	0.577
Managerial learning processes	D1	-0.06616	0.7044	-0.09	0.926
	D2	0.67038	0.7170	0.94	0.355
Requalification of existing staff	D1	-0.62773	0.4469	-1.40	0.167
	D2	-0.30131	0.6956	-0.43	0.667
Constant					
Alpha_1		-0.95391	0.8755	-1.09	0.282
Alpha_2		-0.50157	0.6990	-0.72	0.477
Alpha_3		0.43585	0.7175	0.61	0.547
Alpha_4		1.46882	0.7742	1.90	0.065
Measures of fit:					
Likelihood Ratio Chi-square with 14 d.f., prob= 0.1743:				18.7604	
-2 Log Likelihood for full model:				155.5826	
-2 Log likelihood for restricted model:				174.3430	
Percent Correctly Predicted:				35.0000	

A check on the robustness of the above results was next carried out by focusing on all explanatory variables with significant coefficients from the partial analyses, and carrying out the ordered probit analysis for these variables separately. By starting with all these variables simultaneously and eliminating these ones by one on the basis of their degree of insignificance (the sequential elimination procedure advocated by Theil 1971), it appeared that the hardware-related factor of climatic conditions remained significant in all these model runs, while the other factors tended to become statistically insignificant.

Table 9.6. Probit results for ecoware factors

Variable		Probit Estimate	Std. Error	t-value	p>|t|
Visual disturbance	D1	0.45248	0.5544	0.82	0.419
	D2	0.28094	0.5128	0.55	0.586
Harmony with public space functions	D1	0.93288	0.5594	1.67	0.102
	D2	-0.14986	0.5491	-0.27	0.786
Harmony with building functions	D1	-0.35565	0.5035	-0.71	0.484
	D2	0.26922	0.5275	0.51	0.612
Environmental awareness	D1	-0.10440	0.4729	-0.22	0.826
	D2	0.42067	0.5533	0.76	0.451
Non-tangible social benefits	D1	1.31168	5.5725	0.24	0.815
	D2	1.39684	5.5688	0.25	0.803
Constant					
Alpha_1		0.18545	5.5705	0.03	0.974
Alpha_2		0.63554	5.5674	0.11	0.910
Alpha_3		1.63911	5.5812	0.29	0.770
Alpha_4		2.59579	5.5865	0.46	0.644
Measures of fit:					
Likelihood Ratio Chi-square with 10 d.f., prob= 0.1943:				13.5554	
-2 Log Likelihood for full model:				160.7876	
-2 Log likelihood for restricted model:				174.3430	
Percent Correctly Predicted:				43.3333	

In order to have a further check on the robustness of the above presented results of the ordered probit analyses, we have also applied for the respective cases a binomial logit model on the data. The dependent variable was then classified into a category for negative and neutral scores, and a category for positive scores, respectively (the robustness of the assignment of the neutral score was checked by shifting the neutral category to the category with positive scores). These logit models did not provide additional or more interesting results. A plausible reason is that relevant information in these models is lost due to the ordering of the dependent variable. A major problem inherent in the use of the discrete data models is apparently the low level of measurement of the success rates of the introduction of new energy technologies in cities. Seen from this perspective, it seems plausible to resort to non-parametric statistical approaches (dealing with categorical data), as embedded in rough set analysis. This will be set out in the next section.

9.4.2 Individual energy technologies: a rough set analysis

In our comparative analysis of urban sustainability initiatives, we will use a meta-analytic approach for identifying key factors influencing the success rate of individual energy saving technologies in cities. Meta-analysis is originally a statistical procedure for combining and comparing research findings from different studies focusing on similar phenomena (see Hedges and Olkin 1985; Light and Pillemer 1984; and Wolf 1986). Meta-analysis is particularly suitable in cases where research outcomes are to be judged or compared (or even transferred to other situations), especially when there are no controlled conditions. Given the comparative purposes of meta-analytic methods, it is clear that the information basis for carrying out a meta-analysis has to be designed with a maximum of uniformity with respect to the case to be compared. In the past decade, a variety of meta-analytical methods has been developed (see e.g. Hunter *et al.* 1982; and Rosenthal 1991). Most meta-analytical techniques are designed for sufficiently large numbers of case studies, so that statistical probability statements can be inferred. In this context, standard regression methods or - in case of categorical data - discrete choice methods are often employed. And in this respect, meta-analysis has clearly demonstrated its validity and usefulness as a methodological tool for comparative study in the social sciences. Recent applications in the field of transportation science and environmental science can be found in Button and van den Bergh 1997, Button and Kerr 1996, Button and Nijkamp 1997, Nijkamp and Ursem 1998b while a broad overview can be found in van den Bergh *et al* 1997.

Meta-analysis is essentially a mode of thinking and may comprise a multiplicity of different methods and techniques, which are often statistical in nature. Especially in case of quasi-controlled or non-controlled comparative experimentation, the level of information is often not cardinal, but imprecise (e.g. nominal, qualitative, fuzzy). In recent years, *rough set* theory has emerged as a powerful analytical tool for dealing with 'soft' data. Rough set theory aims to offer a classification of data (measured on any information level) by manipulating these data in such a way that a range of consistent and feasible cause-effect relationships can be identified, while it is also able to eliminate redundant information (see for more details also Pawlak 1991;

Slowinski and Stefanowski 1994; Greco *et al.* 1995; and Matarazzo and Nijkamp 1997).

Rough set analysis is an exploratory, non-parametric statistical method that is able to handle a rather diverse and less directly tangible set of factors in a decision-theoretic context, normally in the form of 'if...then' statements. Rough set analysis, proposed in the early 1980's by Pawlak (1982; 1991), provides a formal tool for transforming a qualitative data set, such as a collection of past examples or a record of experience, into structured knowledge, in the sense that it has the ability to classify objects in distinct classes of attributes (Van den Bergh *et al.* 1997). In such an approach, it is not always possible to distinguish objects precisely on the basis of given numerical information (descriptors) about them. This imperfect information causes indiscernibility of objects through the values of the attributes describing them and prevents them from being unambiguously assigned to a given single class or category. In this case, the only sets which can be precisely characterized in terms of values of ranges of such attributes are lower and upper approximations of the set of objects. Rough set theory has proven to be a useful tool for a large class of such qualitative or fuzzy multi-attribute decision problems. It can effectively deal with problems of explanation and prescription of a decision situation where knowledge is imperfect. It is helpful in evaluating the importance of particular attributes and eliminating redundant ones from a decision table, and it may generate sorting and choice rules using only the remaining attributes so as to identify new choice options in policy problems and next to rank them. An extensive empirical application of rough set theory to a comparative study on regional tourist multipliers can be found in Baayens and Nijkamp (1998).

It is clear from the various recent applications of rough set theory, that it is essentially a classification method, devised for non-stochastic information. This also means that ordinal or categorical information can easily be taken into consideration. This makes rough set analysis particularly useful as a meta-analytical tool in case of incomplete, imprecise or fuzzy information on phenomena to be compared. In the context of our evaluation of sustainable city initiatives, a rough set analysis applied to our data set may be useful for the following purposes:

- assessment of the significance of particular condition attributes;
- construction of a minimal subset of variables ensuring the same quality of description as the whole set, i.e. the reducts of the set of attributes;
- intersection of those reducts giving a core of attributes that cannot be eliminated without disturbing the quality of description of the set of attributes;
- elimination of irrelevant attributes.

The use of rough set analysis for the search of explanatory factors for the success rate of urban energy policy options in our study employs two successive steps:

Information survey. In our case the information survey is derived from the ordinal score data discussed previously. In our approach we distinguish five dependent attributes, namely the main Pentagon groups. For each group of Pentagon factors, an average value is calculated from the scores of the elements belonging to this group. For the rough set analysis we use - in contrast to the statistical analysis - a subsample of the expert interview data, namely only data on the expert opinions regarding existing applications of renewable energies. This concerns useful information for each individual technology; for each technology we group the interview cases that represent a real implementation of a particular technology in a given city.

Classification of information. Next, all observations are classified into various categories for each attribute separately. This applies to both categorical and ratio information. In general, some sensitivity analysis on the classification used is meaningful, as a balance has to be found between homogeneity and class size. This classification exercise leads then to a so-called decision table, in which all observations on urban sustainability initiatives are subdivided into distinct classes for each relevant attribute.

In our case, only the independent variables have to be classified (as these are continuous variables given the averaging procedure). As indicated in the previous section, the dependent variable is already of a discrete nature, with values ranging from 1 to 5. Any further classification of the dependent variable would lead to a loss of information. The classes created here for the independent variables are the following (see Table 9.7):

Table 9.7. Classification of independent variables

Class no.	Range in which mean score value falls
Class 1	[0; 3]
Class 2	<3; 3.5]
Class 3	<3.5; 4]
Class 4	<4; 5]

Applying this classification of the original data, a set of coded variables results for each of the subsamples of the individual technologies. The technologies taken into consideration are: photovoltaic, solar water heating, collective wind energy, refuse biogas and CHP/DH applications, as only for these applications two or more observations exist. For each of these technologies, four main sets of relevant performance indicators and outputs can be calculated: the *reducts*, the *core*, the *lower and upper approximation* and the *accuracy* and *quality* of the classification. These will now be described and interpreted.

We will first pay attention to the *reducts*, i.e. all combinations of explanatory or independent variables which can completely determine (or explain) the variation in the dependent variable, without needing other explanatory variables. For all sub-samples, the reducts are given in Table 9.8. A first conclusion to be drawn from Table 9.8 is that both photovoltaic and CHP/DH energy options have only one minimal set.

Table 9.8. Results of rough set analyses for subsamples of each renewable energy policy

Photovoltaic systems	
Core	{hardware, software, ecoware}
Minimal sets	{hardware, software, ecoware}
Solar water heating	
Core	-
Minimal sets	{finware, ecoware}
	{hardware, ecoware}
	{software, finware, orgware}
Collective wind energy	
Core	{all elements}
Minimal sets	{all elements}
Refuse biogas	
Core	-
Minimal sets	{hardware}
	{finware}
	{orgware}
	{ecoware}
CHP/DH	
Core	{hardware, orgware, ecoware}
Minimal sets	{hardware, orgware, ecoware}

This means that in both of these cases only one combination of variables can explain the variation in the perceived success of the energy option. Hardware and ecoware are common variables in both cases, and have therefore a strong explanatory power for the success rate of the technology. This result confirms the finding from the general statistical analysis that hardware is a dominant explanatory factor. Additional variables in the reduct are software in the case of photovolatic systems, and orgware in the case of CHP/DH systems.

In the case of solar water heating there are three competitive explanatory frameworks. Hardware and ecoware aspects are variables which occur in two of the reducts, and therefore here again have a dominant character. For collective wind energy and refuse biogas, the situation is less clear. Collective wind energy does not show a particular important explanatory variable, as all variables are present in the reduct. In the case of refuse biogas, there are only two observations and therefore the set of perceived success rates (3 and 4) is discernible on the basis of almost each single pentagon factor.

It is interesting to see also that the factor ecoware is the one mostly occurring in the reducts across all technologies. This confirms the assumption that the success of a local application is strongly dependent on the specific (geomorphological and building) characteristics related to the site itself. Exceptions are collective wind energy and refuse biogas (no information available for refuse incineration, CHP/R and energy use saving). This is quite plausible as these latter two applications need not to be integrated with the built environment (locations outside the urban built area are preferred for these applications); specific city characteristics will then not particularly contribute to their success.

Next, we discuss the *core*, i.e. the set of variables that are in all reducts as discussed under (1), or that are part of all competitive explanations. Only photovoltaic, collective wind energy and CHP/DH options have a non-empty core. Because in these three cases there is only one reduct, the core is equal to these reducts. In the case of each of these technologies, the variables in these reducts are essential (as well as a complete set of) factors in explaining the success rate variation.

Finally, we will take a look at the third and fourth factor. We have to assess then the *lower and upper approximation* and derived accuracy of relationships for each value class of the decisional variable. The latter is the lower divided by the upper approximation of each class. From this information we can calculate the *accuracy* and *quality* of the classification (i.e. in relation to the choice of thresholds). In various cases in our application, these indicators are less than 1 (which is the maximum value). In the cases of solar water heating, refuse biogas and CHP/DH systems, both the accuracy of the classification and the quality of the classification appear to have the highest possible value 1, so that the robustness and reliability of the results from the rough set analysis for these latter technologies leaves nothing to be desired.

The overall conclusion that we can draw from the analysis of the individual technologies is, that, across the technologies, the hardware and ecoware components have relatively the strongest influence on the success rate of the technologies in comparison with the other factors software, finware and orgware. This result confirms the finding from the ordered probit analysis that hardware is a dominant explanatory factor. This confirms also the assumption that the success of a local application is strongly dependent on the specific (geomorphological and building) characteristics related to the site itself. Exceptions are collective wind energy and refuse biogas (no information available for refuse incineration, CHP/R and energy use saving). This is quite plausible as these latter two applications need not to be integrated with the built environment (locations outside the urban built area are preferred for these applications). Specific city characteristics will then not particularly contribute to their success.

Between the different technologies there still seem to be important differences in the influence of these factors. For this reason the software component seems to be an additional important success determinant in the case of photovoltaic systems, and orgware in the case of CHP/DH systems. The latter finding can be explained by the fact that many actors and user groups are involved with the implementation of a CHP/DH system in a city, which perhaps results in relatively stronger requirements on the organizational structure. Only with good agreements and close co-operation between the different actors, may such a system operate successfully.

9.4.3 Conclusions

A statistical analysis of potentially influencing factors on the success rate of renewable energy technologies, applied to both existing and future (hypothetical) situations, resulted in two key forces with a significant influence, viz. hardware aspects and financial aspects. More detailed aspects that seem to have been of influence were: local climatic conditions (affecting the efficiency of a given technology), the participation of large-scale users, the profitability of suppliers[1],

[1] This result confirms the importance of the supplier's perspective in the adoption of these technologies, as emphasized extensively in Part B.

external energy market conditions and the harmony or compatibility of the technology with the locality.

The rough set analysis, applied to existing implementations of renewable energy technologies, provided also some interesting conclusions. Hardware factors - as a necessary though not sufficient condition - appeared to have also in this analysis a relatively strong influencing power on the success rate of the relevant technologies. In addition, this analysis suggested that the geographical circumstances, at the local level, are relatively strong explanatory factors, sometimes stronger than financial or organizational factors. Thus, the rough set analysis confirms the main result of the statistical analysis on the influence of hardware factors, but puts in addition more emphasis on the role of site-specific conditions.

It is also interesting to see that financial aspects seem to be envisaged as more important when virtual cases are included in the sample. This indicates that especially with regard to investment decisions the financial aspects may play an important role, and that in later phases (after the financing of the project) the success of the technologies is - in the eyes of the experts - more dependent on other factors.

Between the different technologies there seem to be important differences in the influence of these factors. For example, the software component seems to be an additional important success determinant in the case of photovoltaic systems, and orgware in the case of CHP/DH systems. The latter finding can be explained by the fact that many actors and user groups are involved with the implementation of a CHP/DH system in a city, which perhaps results in relatively stronger requirements on the organizational structure. Only with good agreements and close co-operation between the different actors, can such a system operate successfully.

Given the result that in addition to technological factors, the spatial differences are clearly important for the success of sustainable city policies, it is worthwhile to carry out further research on this issue. The technique of meta-analysis can be used for assessing which cities are likely to be successful in their energy policies, and whether and why such cities are more sustainable than others. For the purpose of comparative policy evaluation, it would of course be most important if we would identify which form of renewable energy is best suitable in which type of city. Therefore, not only factors related to the application of the particular technologies should be taken into consideration, but also socio-economic and morphological characteristics of cities. It would certainly be interesting to carry out more meta-analytic research which concentrates on such characteristics of localities in the context of urban ecology, based on a much larger sample.

10 Success Factors and Barriers of Urban Energy Policies: Interdependencies among Different Critical Factors

10.1 Introduction

We will in this chapter provide a synthesis of the benefits of renewable energies and the aspects acting as critical conditions for their adoption, based on the field work in Chapters 8 and 9. To understand the combined impact of the distinguished critical success factors, we have to take into account that these factors may show interdependencies. Therefore, we will also elaborate on these interactive aspects.

In Section 10.2 we will provide a synthesis of the qualitative and quantitative analyses, in which also the influence of some specific factors related to the urban built environment and population are summarized. In Section 10.3 an elaboration of the internal dependencies between critical success factors is presented. Finally, in Section 10.4 we will set out some final conclusions.

10.2 Synthesis of Qualitative and Quantitative Results

In this section we will provide a synthesis of the qualitative and quantitative analyses carried out in Chapters 8 and 9, with a comparison of the different renewable energy options. We will set out the impacts on urban sustainability and influence of the main groups of critical success factors (*hardware, software, finware, orgware and ecoware*). This overall picture will be further detailed for the separate critical success factors by each group of factors.

To start with the basic influence of the urban built environment and population characteristics, an overview of this impact as it emerges from the qualitative analysis is shown in Table 10.1. There seem to be some important differences between renewable energy technologies which make their potential success dependent on certain specific urban properties. First of all, we may mention the difference between PV/SWH systems and CHP/DH systems in their infrastructural costs. These costs are in the case of CHP/DH relatively high and are mainly borne by the investors, mostly (privatized) energy companies. This means that they will only invest if energy demand is high enough and spatially concentrated to reach a sound cost/revenue ratio. Therefore in the case of CHP/DH a high density of population and houses, and a high energy use per capita (and thus, a low insulation, or older building types) are favourable conditions. PV and SWH systems on the contrary require in general low population and house densities, due to the fact that a spatially dispersion of buildings (or at least, non-stacked housing buildings) is necessary for the instalment of collector plates.

Table 10.1. Relationships between urban characteristics and the potential of renewable energy policies

Urban characteristic	PV	SWH	Wind	DH
Population density	-	-		+
Income per capita	+	+	+	
Dwelling density	-	-		+
Dwelling age				+
Dwelling ownership	+	+		
Energy use per capita				+

Note: - negative influence; + positive influence.

Table 10.2 is an overall cross-evaluation for renewable energy technologies. In this table all available quantitative information, as it results from the various analyses on factor scores (cases and non-cases taken together), is shown in a condensed form per renewable energy policy option. It must be stated that the quantitative information in this table is indicative only, and therefore great caution should be taken with its interpretation. Nevertheless, we can draw some general conclusions on the differences between the technologies.

First, the technologies which have a relatively large contribution to the reduction of pollution (photovoltaic, solar water heating and wind energy) are also the ones for which financing is a major bottleneck. For the technologies which are strong in terms of their hardware (refuse incineration, refuse biogas and CHP applications), the financial problems are smaller than with other technologies, but also show smaller reductions of air pollution.

Second, it appears that the technology of wind energy faces the largest bottlenecks in comparison with the other technologies. Wind energy scores even lower than PV technology with regard to software and ecoware. An intuitive explanation of this could be found in its often mentioned disadvantages: the consequence of its fluctuating energy generation for the user and its infringe on landscape and living surroundings.

In Tables 10.3 and 10.4 more detailed summarized information is presented on separate critical success factors. These two tables are structured according to the framework provided in Table 3.1. Table 10.3 regards the supply side of the technologies and Table 10.4 the demand side.

An effort in classifying the technologies according to their potential success rate would then start with making a distinction between two groups of renewable energy technologies, given the clear differences amongst these.

On the one hand, we may consider the relatively consolidated district heating and combined power generation and all technologies regarding the use of refuse, which has been implemented recently on quite large scales. This group is supposed to have a considerable further development rate in the future, in spite of the delays in the past, taking into account that the urgency and the complexity of the environmental issues have made them essential.

Table 10.2. Potentials, bottleneck factors and determinants of success rates: comparison of renewable energy technologies.

Renewable energy option	Contribution urban sustainability		Level of fulfilled critical success factors					Determinants of success rate
	Saving of non-renewable sources	Pollution reduction	Hardware	Software	Finware	Orgware	Ecoware	
Photovoltaic	+/++	+++	••	••	•	•	••	hardware software ecoware
Solar water heating	+	+++	••	•	•	••	••	finware ecoware
Wind energy	+	+++	••	•	•	•	•	non specific
Refuse incineration	+++	+	•••	n.a	••	•	••	-
Refuse biogas	+	+	•••	n.a	••	••	••	no specific
DH	++	++	•••	•••	••	••	•••	hardware orgware ecoware

Notes:

n.a. not applicable

+, ++, +++ contribution respectively low, moderate and high in comparison with other technologies

•, ••, ••• the level to which the factor acts as bottleneck in comparison with other technologies is respectively high, moderate and low.

Table 10.3. Critical success factors on the supply side for the adoption process of innovative energy production technologies

Pentagon factor	TECHNOLOGIES		
	Photovoltaic system	**Solar water heating**	**Aeolian systems**
Techno-logical	• R&D is very costly *while further development has been slowed down in recent years*: innovations are necessary on the silicon extraction (o) • configuration of the technology is relatively simple (n) • heart of the technologies (like inverters) are less efficient (o) • non-favourable climatic conditions obstruct the success of many initiatives (o) • for having an efficient connection to the electric network, PV systems need to have a size that is at present too costly (o) • *possible synergies are to be exploited in a domestic and independent context (n)*	• more efficient cost/energy production ratio than other house-based technologies (n) • technology is simple (n) • standards are being developed (n) • high compatibility with existing housing energy systems (n)	• small size is efficient only in an isolated configuration; large sizes have to be connected to the network (o) • technology costs depend on the wind speed (in general very low) (o) • technology is quite simple (n)

Note: in this table italic sentences mean an analysis in a dynamic setting. The signs n and o mean that the critical factor is respectively a success factor or a bottleneck in most of the case cities investigated.

Table 10.3 (continued).

Pentagon factor	TECHNOLOGIES		
	Photovoltaic systems	**Solar water heating**	**Aeolian systems**
Financial	• sunk costs of appliances (combined with the limited efficiency) (o) • many cases national subsidy schemes for energy saving in buildings are perceived not sufficient, given the costs (o) • low costs of maintenance (o) • very low profitability: pay-back periods are long (60-80 years) (o) • financing involving users is not proposable given the non-profitability (o) • prices of other energy sources are more competitive (o)	• existing profitability to producers (n) • large-scale projects might have only moderate losses for the suppliers (e.g. the case of Gouda, where no governmental subsidies have been needed) (n)	• sunk costs of new plants (o) • high initial investment costs for land, aerogenerators, and the connection to the network (o) • low costs of maintenance and of technical re-qualification (n) • in many cases governmental financial support does not repay the initial expenses and investments (decreasing with the increase of the wind) (o)

On the other hand, we can group together the PV and the aeolian technologies. Although the technological state of these systems is becoming more and more mature, the initial costs remain relatively high and their energy efficiency very low, because of the indivisibilities linked to their implementation scale (see Part B). These indivisibilities grow when the planned switch from a small size to a large size is fast. For this reason, it is at present hard to imagine a promising future development of these initiatives. In addition, there are geographical features that are in many cases prohibitive to these systems, as seen in many of the case cities.

Table 10.3 (continued).

Pentagon factor	**TECHNOLOGIES**		
	Pothovoltaic systems	**Solar water heating**	**Aeolian systems**
Organiza-tional	• Strong political-organizational willingness of public operators is necessary to bear the high costs (o) • the market is not free: there are only few operators carrying out R&D, few technology producers, few and non- competent installators (o) • low costs of management (due to the simplicity of the system) (n) • *a present wide-scale introduction is unlikely due to the high costs (o)* • the willingness of public operators to manage the plants c/o the householders is often lacking (o)	• availability of marketing experience to sell the product (n) • the market of other energy sources has still more competitive prices (o)	• the market of other energy sources has much more competitive prices (o) • the fact that in many cases windmills are not implemented on large scales represents at present an appropriate step (n) • low costs of management (n)
Ecological /Social	• saving of non renewable energy sources (n)	• saving of non renewable energy sources (n)	• substitution of non-renewable energy with cleaner sources (n)

Note: in this table italic sentences mean an analysis in a dynamic setting. The signs n and o mean that the critical factor is respectively a success factor or a bottleneck in most of the case cities investigated.

In the course of our analysis we have taken into account this above mentioned difference. In fact, it is true that the technology is in general easily available, and the reliance on foreign experience is not perceived as a negative element, unless international technological problems exist. But it is just in the case of an early introduction that the greatest costs can emerge in terms of know-how acquisition given the absence of a widespread knowledge in the field of renewable technologies.

Table 10.3 (continued).

Pentagon factor	TECHNOLOGIES	
	Biogas from disposals	**Incineration of waste**
Techno-logical	• *R&D has been costly but now has achieved a good level (n)* • technology is widely available and implemented (n) • technical compatibility with existing disposal plants (n) • availability of related technologies (e.g. interface with electric network) (n) • *level of competence to use the technology when waste operators and energy producers are not coincident (o)* • *sunk costs can emerge when using more powerful engines; this raises the question how the use of biogas can be increased (o)* • strategic role for integrated waste management (n)	• local operators are needed for local sources that can be lacking when the sources are too specific (at the spatial micro level) (o) • particular climatic conditions can affect the incinerator characteristics (with costly environmental and technical consequences) (o) • resort to foreign experiences (n) • technical incompatibility with regard to variable composition of waste (o) • need for specific know-how *(waste operators often lack experience/know-how to produce energy without know-how of energy companies) (o)* • *economic advantages for industrial activities operating without an own thermal plant (n)*

Note: in this table italic sentences mean an analysis in a dynamic setting. The signs n and o mean that the critical factor is respectively a success factor or a bottleneck in most of the case cities investigated.

Table 10.3 (continued).

Pentagon factor	**TECHNOLOGIES**	
	Biogas from disposals	**Incineration of waste**
Financial	• pay back periods are relatively short (n) • *at present in many cases, initiative have time-limited subsidies and the question is whether these initiatives have social viability afterwards (o)* • high initial location costs (o)	• high costs of R&D (o) • high sunk costs when old and unsafe incinerators exist, which have to be substituted (o) • high costs of incinerators (o) • resort to governmental subsidies (n) • relatively short pay back periods, especially when in cases of private management (n) • resort to tools of third part financing (n) • incinerators are not efficient enough and have a lower cost/benefit ratio when they are not connected to a large-scale or small district heating system (o)

Note: in this table italic sentences mean an analysis in a dynamic setting. The signs n and o mean that the critical factor is respectively a success factor or a bottleneck in most of the case cities investigated.

In many cases however, political and organizational factors - more than the availability of technology - turned out to be the main reason for the delay in the adoption processes. An integrated supply system of energy and connected services in the city (for example supply of different energy vectors and waste management by an integrated energy company) seems to be, in effect, necessary to assure the minimisation of costs in terms of competence acquisition.

Another area in which very important bottlenecks can be found is the financial field. In this case the time variable is important, and an interesting paradox can be underlined. Financial instruments did not exist some years ago, but nevertheless some initiatives were born without these instruments thanks to other favourable organizational and market conditions. At present, financial tools exist and not only are they sometimes necessary, but they are also perceived as not sufficient for the future development of other initiatives (like the rapid enlargement of district heating to a very large size, in the case of Turin, or like the production of biogas from disposal).

Table 10.3 (continued).

Pentagon factor	TECHNOLOGIES	
	Biogas from disposals	**Incineration of waste**
Organiza-tional	• organizational cooperation can create problems; it is better not to have too many actors involved in the production-carriage-selling of the energy (o) • no specialisation problems when the management is boarded out to thirds (n) • introduction of this technology coincided with a general situation of waste emergency (n) • synergies with integrated waste management (n) • *when energy producers and waste managers coincide, few problems will emerge (n)* • *delays in the start-up of initiatives (o)* • third management of technical aspects (n)	• there is a need for the involvement of this technology in integrated waste management (o) • employment requalifications (n) • the market of operators is quite narrow, given the high specialisation required (o) • operation often boarded out to thirds (n) • *up to now large delays occurred with the implementation, and further waiting will make the situation more difficult (o)*
Ecological/ Social	• synergies with integrated waste management (n) • source is available at a negative cost (n)	• reduction of refuse in disposal (reduction of methane, bad smells and space problems) (n) • source is available at a negative cost (n) • air pollution due to incineration (o)

Note: in this table italic sentences mean an analysis in a dynamic setting. The signs n and o mean that the critical factor is respectively a success factor or a bottleneck in most of the case cities investigated.

Table 10.3 (continued).

Pentagon factor	**TECHNOLOGIES**	
	DH with non renewable sources	**DH with renewable sources**
• **Techno-logical**	• high (European) level of R&D given the information flows and the high standard required (n) • consolidated and widely available technology (n) • *the achievement of this level has taken a long time period (o)* • technical compatibility with old and obsolete thermal plants and old water network (n) • technical compatibility with gas/ oil network (n) • large strategic role in the energy supply system (n) • reliable technology (n) • flexibility of plants (use of multiple fuels) (n) • it is impossible to switch from methane to DH (coexistence is better) (o) • *a large implementation size is an obstacle if it has to be achieved within a short time period (o)* • *it is easier to begin later to exploit experiences in other sites (n)*	• technology is easily available especially at the international level (n) • technical compatibility between DH and gasoil (n) • the variable composition of refuse causes unsafe performances and stronger safety parameters are needed (o) • sufficiently high level of R&D (n) • technological costs of depuration (n)

Note: in this table italic sentences mean an analysis in a dynamic setting. The signs n and o mean that the critical factor is respectively a success factor or a bottleneck in most of the case cities investigated.

Table 10.3 (continued).

Pentagon factor	**TECHNOLOGIES**	
	DH with non renewable sources	**DH with renewable sources**
Financial	• *high R&D costs (o)* • no sunk costs of the old technology (n) • self financing initiatives (although in many cases still public financing is exploited) (n) • public actors can bear long pay-back periods (n) • high initial investment costs (n) • low maintenance costs (n)	• incinerator makes up a very high share of total initial costs (o) • no sunk costs of the old technology (n) • profitability is more at risk than DH with a non-renewable source, due to the high initial costs (o) • subsidy schemes are only available at the national level and not from the EU, because of the 'non- innovative' character of the incinerator (o) • *possibility to resort to the tool of third part financing (n)*

Note: in this table italic sentences mean an analysis in a dynamic setting. The signs n and o mean that the critical factor is respectively a success factor or a bottleneck in most of the case cities investigated.

In conclusion, as far as the time patterns of adoption are concerned, something can be said. First of all, not always does a non-introduction represent a failure: the non- adoption of initiatives regarding PV and aeolian energy on a large scale for example, is at present the best choice to be made in the cities investigated. Even if these technologies are the cleanest and the most simple technologies that can be applied, they do not seem to form, now and in the near future, the basis on which the control of environmental pollution and the energy use policy of the cities can be founded.

Table 10.3 (continued).

Pentagon factors	TECHNOLOGIES	
	DH with non-renewable sources	**DH with renewable sources**
Organiza-tional	• integrated energy companies facilitate the peaceful coexistence between DH and methane (n) • technical requalification costs (o) • organizational - political cooperation (with some exceptions) to face high investment costs (n) • favourable market conditions when DH and methane are not competitive (n/o) • *for large-scale implementations, learning processes and staff re-qualification are costly at the start (o)* • *favourable conditions have been exploited in the past that at present exist no longer (high level of inflation and expensive energy sources): a present introduction becomes more difficult (o)*	• large commitments for an organizational cooperation are needed (even more than in the 'traditional' DH), due to the refuse as energy source (o) • initial bargaining costs (o) • *delays are costly especially considering the waste emergency, as the selling off in its own disposal or, worse, in other cities, has cost too much (o)* • *technical re-qualification costs (o)* • *high learning costs at the start (o)*
Ecological /Social	• large energy saving (n) • reduction of air pollution (n)	• initial location costs and technical studies for environmental impacts (o) • energy saving (n) • source widely available (n) • air pollution from incineration (o) • reduction of refuse in disposal (n)

Note: in this table italic sentences mean an analysis in a dynamic setting. The signs n and o mean that the critical factor is respectively a success factor or a bottleneck in most of the case cities investigated.

Table 10.4. Critical success factors on the demand side for the adoption process of innovative energy production technologies

Pentagon factor	TECHNOLOGIES		
	Photovoltaic systems	**Solar water heating**	**Aeolian systems**
User-related	• the critical mass of users is difficult to be achieved (panels on roofs or walls discourage many) (o) • high technical reliability for users that have panels to be connected to the electric network installed on roofs or walls (n) • technical adaptation is easy (furthermore the maintenance is easy and can be managed by thirds) (n)	• high costs of marketing management (o) • critical mass of users can be achieved through large-scale implementation in new building blocks (n) • low technical reliability (at least in starting years) (o) • role of large-scale users limited (o)	• the system reliability is set off by the discontinuity of the source (the more discontinuity, the lower the number of potential selfproducers) (o) • for users who are connected to the electric network nothing changes (n) • for selfproducing users (connected or not to the network) much changes (n)

The cases of district heating and technologies regarding the energy use of refuse is different. The delays that have occurred until now have caused excessively high costs in terms of monetary and energy losses and the situation of the refuse excess has become unsustainable. It would not be wise to wait any longer, in spite of the public opposition of the population, especially when refuse is involved. It is for this reason (and at least it has become clear also for policy makers and experts) that projects and large-scale adoption of this kind of initiatives cannot be planned without taking into consideration consensus and interests of all the passive and active actors.

Table 10.4 (continued).

Pentagon factor	**TECHNOLOGIES**		
	Photovoltaic systems	**Solar water heating**	**Aeolian systems**
Financial	• sunk costs of new technologies (o) • very low profitability for small or large scale users due to the high costs (o) • the adaptation in terms of costs is not feasible (o)	• with moderate subsidies end-users could have a break-even of energy costs (n) • relative low sunk costs of the technology (n)	• sunk costs of the aerogenerators (o) • *very low profitability (especially for potential individual selfproducers) (o)*
Ecological /Social	• absence of dislike from the environmental point of view by the public (n) • no visibility impacts thanks to panels installed ex-ante (n) • visibility impacts and space problems for large size applications (more efficient) (o)	• house-interior visual problems due to the installation of the equipment (o) • the general environmental awareness can be raised quite rapidly through the large-scale implementation at the household level (n)	• absence of dislike from the environmental point of view by the public (n) • space often not available for aerogenerators parks, because these need to be large in order to be efficient (o) • danger for birds (o) • noise (o) • electromagnetic interferences (o)

Note: italic sentences mean an analysis in a dynamic setting. The signs n and o mean that the critical factor is a success factor or a bottleneck for most of the cases investigated.

Table 10.4 (continued).

Pentagon factor	TECHNOLOGIES	
	Biogas from disposals	**Incineration of waste**
User-related	• the success of this technology is co-determined by the attitude of the public toward separated waste collection and their willingness to change their habits (n)	• *optimal reliability for users, especially when they are connected to a DH system (some plants have such a connection, others not)(o/n)* • potential existence of users that can be connected to small-scale DH systems linked to waste incinerators (industrial users but also other users like schools) (n) • except for a positive change, namely less maintenance, nothing changes for users, with or without DH (connection to an electricity network or even to a DH system does not imply a significant change in their habits) (n) • many potential large-scale users do at present not seem to be interested (o)
Financial	• separated waste collection may have no financial implications for the public (n)	• if the incinerator is not connected to a DH system the possibility of exploiting the advantages for the users (facilitated heat tariffs) goes lost (o)
Ecological /Social	• reduction of danger stemming from disposals (more controls which are more accurate) (n) • reduction of bad smells (from free biogas) (n) • the advantages, even in terms of reduction of bad smells, are not yet perceived as very important by the public (o) • disposal treating activities are always irritating, with or without the recover of biogas (o)	• dislike for the general concept of waste incinerators (o) • prevalent interest for economic factors (if a connection with DH exists, users are attracted) (o)

Note: italic sentences mean an analysis in a dynamic setting. The signs n and o mean that the critical factor is a success factor or a bottleneck for most of the cases investigated.

Table 10.4 (continued).

Pentagon factor	**TECHNOLOGIES**	
	DH with non renewable sources	**DH with renewable sources**
User-related	• high costs of information campaigns (o) • existence of a small niche that can be served with efficiency by the DH system (large users or high dwelling density) (n) • technical compatibility between gasoil and DH is positive also for the users (n) • resistance by users to modify their habits for what concerns the timing of energy supply (some companies would prefer a continuous 24-hour supply instead of the usual 12 hours, but this would be only useful for users that at present exploit night tariffs) (o) • high technical reliability for users (n)	• high costs of information campaigns (o) • existence of a small niche that can be served with efficiency by the DH system (large users or high dwelling density) (n) • technical compatibility between gasoil and DH is positive also for the users (n) • high technical reliability for users (n)
Financial	• absence of sunk costs for users because they usually switch to DH when their gasoil plant is obsolete (n) • initial non-financial participation (no connection costs) (n) • profitability for users (often lower heat tariffs) (n) • *in cases that the initial costs are higher (for large-scale projects), an initial financial participation will be necessary (o)* • no adaptation costs because there is either no change or a positive change (centralization and minor upkeeping) (n)	• absence of sunk costs for users because they usually switch to the DH when their gasoil plant are obsolete (n) • the financial participation of users (for instance, the connection) is more necessary then in the case of traditional DH (costs are higher due to the presence of the incinerator and the depuration appliances) (o) • no adaptation costs because there is either no change or a positive change (centralization and minor upkeeping) (n)

Note: italic sentences mean an analysis in a dynamic setting. The signs n and o mean that the critical factor is a success factor or a bottleneck for most of the cases investigated.

Table 10.4 (continued).

Pentagon factors	TECHNOLOGIES	
	DH with non renewable sources	**DH with renewable sources**
Ecological/social	• location problems in the case of central plants (o) • large territorial impacts of large-scale DH networks (o) • the visibility and other impacts are accepted when there are economic advantages (n)	• large dislike for the incinerators (Nimby syndrome) (o)

Note: italic sentences mean an analysis in a dynamic setting. The signs n and o mean that the critical factor is a success factor or a bottleneck for most of the cases investigated.

10.3 Interdependencies Between Critical Success Factors

The political-organizational willingness seems to be necessary for large-scale activities involving public actors due to their specific responsibilities and goals. On the contrary, it emerged that small size initiatives can be started up also by private actors, induced by a sufficient level of profitability, or moreover by public actors that use them (third part financing). In this case a good level of technological know-how is sufficient (own know-how or coming from outside).

Organizational factors appear to be fundamental especially for two of the technologies in the analysis, namely district heating and the use of biomass. The clear determination of the competent actors involved, their cooperation and their continuous exchange of know-how is sometimes not sufficient. On the contrary, the presence of a single actor could be very efficient, if such an actor is historically linked to the supply of the most important energy sources and to manage the waste system. In this way a peaceful coexistence between different but equally important energy sources for a city (district heating and methane) can be facilitated and prove useful for the actors that accept and will have to accept the challenge of producing energy by using refuse.

Further, it must be noted that for those initiatives which will be exploited on a large scale in the future, efforts have to be made for involving those parts of the population that are affected by negative socio/ecological impacts of these technologies.

As far as the financial variable is concerned, a kind of a paradox exists due to the fact that in the past energy prices have caused more favourable conditions for innovative energy policies than at present. Many initiatives arose when public financial facilities were not yet provided thanks to favourable conditions and the possibility of self financing. Nowadays this would be more difficult due to the lack of these favourable conditions for the initiatives of district heating and in general for those technologies involving the use of refuse. Financing tools necessarily need to be strengthened and made more widely available so that the initiatives stemming from these facilities do not suffer delays or interruptions.

This latter consideration underlines that there is a strong relationship between *financial* and *organizational* aspects. In fact, financial aspects receive more weight when the organisation of energy supply in cities is less clear, or when the conditions of the energy market are such as to create a harmful competition between different sources. In these cases the profitability of initiatives can be lower and consequently, the financial participation of users or more public financing could be required.

On the other hand, not always are long pay-back periods considered as negative financial factors. They are coherent in those initiatives involving a large quantity of financial, technological and economic resources (this is the case of district heating), and can be faced by an actor that has to be a strongly organised public actor.

Another consideration stated previously underlines a further linkage between the *organizational* field and the *economic/financial* field, namely the adaptation of users to pass to new technologies. In general it can be said that the process of user adaptation depends on the reliability (often relatively high) and is faster when the initiatives considered imply no change from the technical point of view. However, the principal factor that makes the user inclined to the switch is the economic consequence that prevails over the environmental aspect. In other words, when economic and financial advantages exist for users - and especially if they are made evident to them by means of information campaigns - some eventual environmental disadvantages will be accepted. But the many environmental advantages related to the introduction of renewable technologies are not sufficient if economic advantages do not exist or are not emphasized.

Also the obvious relationship between *organizational* and *technological* aspects has to be noted. If the organization that has been in charge of energy production and the organization that has only managed refuse, are not the same body, this has important consequences for the technological performance. This gap can contribute - and at present it does in many cases - to keeping the level of necessary technological know-how low for newcomers in the field of energy recovery from refuse. The consequences can be negative, also in terms of costly calls on technical re-qualification or on a third, specialized party.

Another linkage that needs to be emphasized is the relationship between the *technological* field and the *economic/financial* field. First, the technical compatibility that often occurs between old and new technologies (the most important example are district heating applications based on old, obsolete and traditional thermal plants) allows sometimes to avoid part of economic initial costs of the plants and especially to avoid eventual sunk costs of the old technology. On the contrary, for initiatives like photovoltaic and aeolian systems, high costs cannot (yet) be avoided. The technological level in these cases keeps the costs of plants (PV modules and aerogenerators) too high, and the profitability too low. Long pay-back periods are unavoidable due to the difficulty of repaying the initial investments. Secondly, it should not be forgotten that when the level of technological know-how is not sufficiently high, more reliance on foreign experiences becomes necessary. In that case, additional costs in terms of 'competence acquisition' occur which can be a substitute for internal R&D costs (special scientific staff).

Finally, with regard to the relationship involving *environmental* aspects, it may be concluded that location problems have important implications on economic and organizational variables. Location problems imply a large effort for studying the environmental impacts in search of the optimal place and of a solution of social conflicts stemming from choices which are often not accepted by citizens. On the other hand, a lack of space needed for large sized plants (in some cases the only way to be efficient in terms of energy saving) has to be taken into consideration because it has far reaching consequences from an economic point of view.

10.4 General Conclusions

A final policy recommendation for future urban energy projects can be set out in four main lines.

Firstly, the relevant innovative energy technology to be used should preferably be available and already be applied, so that its transition to the new project can be a feasible target for the local situation. The state of the technology is indeed important for the success of a project; however, once it has become mature it does not form a barrier anymore. In many cases the technology is imported from other European countries. The local applications can become successful if selected consultants are well experienced and suitable for this task. After an adjustment period, the technological basis fits with the local setting and a sufficient level of know-how is developed.

In the second place, the existence of proper financial schemes which are able to deal with the investment cost is the most crucial factor. The investment costs are high and the risks are also high. In this respect, there is hardly a private investor willing to undertake pilot initiatives. Until now, financial models for pilot projects relied mostly on European programmes and governmental loans.

Thirdly, the establishment of a crucial mass of users emerges as an important element. Once the project has been designed, the provision of suitable information to the public should create a crucial level of demand. The prerequisite for this evolution is that a project is beneficial for the individual user. It seems that projects with clear economic benefits for the user are welcome. But these benefits could be estimated on other bases as well, for instance, flexibility, convenience, clean use etc.

Finally, the project should be economically viable for the producer as well, so that it presents attractive options.

Under the above conditions a successful pilot project might be designed and developed. Once it has been applied successfully, it may create the potential for expansion and for similar initiatives in other places.

Trying to explain the virtual cases, i.e. cases where there are potentials for introducing renewable energy projects but no initiative is undertaken, is a more difficult task. It seems that the crucial factor is the lack of a pioneering actor who could anticipate the potentials and the relevant profits and undertake action. This lack is strongly connected with the existence of a suitable legal framework, at least in the past. The legal basis should permit and promote such initiatives by liberalizing the entrance to the renewable energy production and by creating the proper conditions for a market to function.

The exploration of positive and negative past experiences with cases of renewable energy implementations in order to draw lessons for the future should take into consideration institutional settings and conditions. Advancing initiatives for exploiting renewable energy resources is a relatively innovative and new direction in Europe. The traditional setting was oriented towards social targets, while power generation was an exclusive task of large public bodies. It also did not have the necessary flexibility for the orientation towards exploring the potentials for other applications as well. This framework was supported by a rather strict legislation. The needs of the contemporary society emerging under the pressure of environmental problems and the scarcity of non-renewable resources pressed for alternative modes in energy production. However, the expansion of new technology applications cannot be based on individual initiatives alone. There is a need for designing pilot pioneering projects which will identify the potentials of alternative energy production. The pioneering nature of these projects should concern not only technical conditions, but also financial schemes, organizational structure and environmental and social benefits.

PART D

CONCLUSIONS: POLICY RECOMMENDATIONS AND GUIDELINES

11 Policy Recommendations and Guidelines for Renewable Energy Technologies in Cities

11.1 Urban Sustainability Revisited

The concept of urban sustainability is not a simple panacea for all environmental city concerns . It is a complex and multi-faceted concept which cannot be simply equated with urban environmental policy or green urban movement. The idea of a sustainable city refers to the future of the city, to effective local policy-making at the interface of economic, social and environmental objectives with a view to long-term continuity of the urban area.

In the foregoing chapters it has been illustrated that energy policy is a key issue for urban sustainability. In this context, a decoupling between energy consumption per capita and urban gross domestic product per capita is a prerequisite for long-range survival of the city. Consequently, the *double dividend* principle of energy saving and urban sustainability seems to be the basis for a wise urban policy.

Apart from economic mechanisms involved in the sustainable city concept, social aspects must be considered. And in this framework, the creation of awareness through communication and information campaigns is equally important. Demonstration projects may play a key role in this context. Thus, participatory forms of urban sustainability policy seem to be the best candidates for a successful achievement of urban environmental targets, not only in terms of economic welfare, but also with regard to local quality of life. It goes without saying that high urban environmental quality will also attract new investments, so it is in the interest of the urban economy to invest also in environmental projects. This message has been clearly mentioned in a recent OECD report (OECD 1995): "*Private business as well as public policy therefore need to include environmental considerations, not as marginal items but as* ***deeply integrated concerns*** *at every stage of strategies and* *policies - in the same way that safety and social concerns are now integrated in the functioning of all firms in OECD countries Clean technologies for products and processes affecting cities or being used in cities, must be further developed but many of them are already available and could be quickly adopted provided that social and environmental costs are properly internalized in the market economy. In addition, more environmentally-conscious urban planning and organization should be developed, in order that land and other urban resources be used with more attention given to quality and conservation. In certain fields, like urban energy and transport, that may even mean controlling, or limiting demand.*"

This study has demonstrated that the road towards urban sustainability is paved with many stumbling blocks. Our theoretical analysis has highlighted some of the impediments related to the acceptance of renewable energy technology as a result of institutional forces and market organization. The empirical and policy parts have shown that - despite a great many barriers - various promising initiatives do exist, which can be put into operation and which may offer a significant contribution. However, the exchange of knowledge and information on successful city policies is underdeveloped. The same applies to the comparative study of sustainable city efforts.

The remaining part of this chapter will offer concrete and practical recommendations and guidelines with respect to the implementation of renewable energy technologies, based on our comparative analysis of renewable energy policies in various European cities.. These guidelines, which are formulated as concrete answers to common problems - encountered before or during the actual implementation process - may help cities direct and stimulate new initiatives in the field of urban sustainability policies. The list of problems encountered is based on the results of the European case study presented in this book. Issues have been selected which have emerged as being of particular relevance to the implementation process and the degree of success of renewable energy technologies.

The guidelines and concerns provided for the removal of bottleneck factors are formulated for each of the distinguished renewable energy technologies, viz. PV systems, collective wind energy, solar water heating, (CHP)/district heating, refuse incineration and energy recovery from biogas. The guidelines also refer to the actors who should undertake action.

11.2 Photovoltaic Systems

The configuration of photovoltaic(PV) systems is, in principle, simple. However, the core of the technology (converters) is not efficient

National authorities should, as a first step, liberalize the market of installators and technology producers. In free market conditions it can be expected that technological development of promising products goes faster than under strict regulations. This also applies to the case of the technology of PV systems as well as all complementary technologies.

Presently, the scale at which plants operate efficiently, is costly

Perhaps the implementation process of PV systems should focus in this early phase on small-scale applications, e.g. domestic PV installations which are not necessarily connected to the electricity grid. Such a switch of strategy from the idea of centralized, large scale PV installations should be supported by suppliers, operators and local authorities.

Climatic conditions can act as major bottlenecks to the actual continuation of concrete public initiatives. This is true about Northern European countries, but also about regions in South-Europe, like Northern Italy, where long and foggy winters occur

Operators and local authorities should weigh contingent conditions and operating costs in order to avoid that the start-up of an initiative is not followed by a continuation or broader implementation in the city. Of course, this may imply that start-up activities in the photovoltaics area may be slowed down. However, looking at available financial resources, it opens up the way to investments in other possible technologies that may be more suitable to location-specific climatic circumstances.

In general, R&D activities are costly and do not reach a sufficiently high standard. This creates a postponement strategy among (potential) initiative takers

National authorities, but certainly also system producers, should undertake more efforts to stimulate research in all energy technologies. An insufficient level of R&D activities has a doubly negative impact on the development of the technologies. Not only the technological development of the systems themselves are involved but also the perception of the initiative takers. Low perceptions slow-down pilot implementations, and in turn, the rate of further technological development.

It is difficult to achieve a critical mass of users in the early phases of the implementation process of photovoltaic systems

A critical point about the level at which potential users are attracted by PV systems is the change in user costs. By lowering the costs of PV systems, the potential market of users (as self-producers) could be enlarged. Financing schemes could help here through close cooperation between system suppliers, system operators and local authorities. Another aspect, which seems to be very important, is the visual design of outside components, since PV systems impact most on the visual outlook of domestic premises. Only when sufficient attention is paid to this aspect, can a market segment for these systems be opened up.

Even when technical adaptation is feasible, the costs are often prohibitive

The challenge here is for system producers to lower the investment costs by stimulating technological research (e.g. on silicon). They should also conduct a long term product strategy based on the improvement of market conditions.

Financing schemes exist but they 'drug' the market. PV systems have a low profitability with high sunk costs while other source prices are more competitive

It is clear that just strengthening the financing structure will not be sufficient. As pointed out above, it is absolutely necessary to develop technology in such a way that the costs will be lower. The introduction of PV systems should also be made possible by producers in areas with non-optimal conditions from an climatic and territorial point of view.

The market is not free. There are only a few technology producers, installators (housing constructors do often not have experience in this field) and operators

There is a clear task for national authorities to open up and stimulate the market of potential producers and operators in the PV area. Of course, it will be very difficult to reach a fairly high level of competition, as the installation, operation and maintenance activities are more or less natural tasks of housing constructors and energy companies. In most countries the organizational development of energy companies is counter-directive. More and more companies fuse into larger regional companies and in order to reduce the present monopolistic behaviour of actors in the energy supply sector, it is a basic requirement that the barrier to newcomers should be reduced.

Public operators are not inclined to manage the plants of households in the case of installation on roofs or walls
The reason why public operators can object to managing and maintaining installations on separate, individual houses is that this is a relatively inefficient operation. This is different in the case of a collective installation of multiple PV systems on blocks of houses, where there is a preference for the installation of pilot applications on blocks of houses rather than single house applications. Nevertheless, as long as the competition in the operators' market does not increase, national authorities should force operators to manage and maintain installations on houses. In turn, the operators themselves should guarantee services to manage and maintain household equipment.

There is weak political-organizational willingness to bear the high costs
This general problem statement regarding PV systems should not only apply to a lowering of the costs of investment through enlarging efforts in R&D activities. Equally important is the creation of a higher level of public awareness about these systems and to improve the attitude of the public towards them. This will aid the acceptance of these systems by the public, and facilitate pilot implementations.

Visibility impacts and space problems are relatively large for centralized, large-scale plants
R&D activities should be increased not only in the core of PV technology, but also in specific system components for household installations. It is envisaged that in the long term the small-scale application of PV systems will have the best future chances from the perspective of space occupation and visibility. It would therefore be wises also to emphasize the development of small-size plants.

11.3 Collective Wind Energy

There is a necessity for a complementary energy supply system in order to deal with the discontinuity of energy from windmills due to their dependence on wind conditions
Much of the success of renewable energy technologies depends on the strategy in which the problem of discontinuity is solved and a continuous supply of energy is guaranteed to the client. This is not purely a technical problem. Energy demand also

is not continuous in time (e.g. in cities there are peaks in electricity demand during the day) but shows a varying pattern of use. In some cases it is possible to find system configurations where the temporal patterns of supply and demand of energy are quite well matched. Moreover, there is the possibility that demand is adjusted to the supply of energy: For example, to a minor extent household activities may be matched with peak supplies, but this can only happen when there is a steady daily temporal pattern in the supply of energy. This is more the case however for solar systems (both PV systems and solar water heating systems) than for wind energy systems. It therefore remains necessary to further develop the technology for storage of energy since peaks in the generation of energy by windmills can never be matched with energy demand in the urban area.

Another, more drastic solution would be that local authorities and operators would avoid initiatives in areas with non-optimal wind conditions until the investment costs have lowered sufficiently. By so doing, money could be reserved for other initiatives that would benefit relatively more from local geographical and climatic conditions.

Large-scale windmill parks have to be connected to the electricity network in order to be efficient, but this is very costly

Until now most pilot projects of wind energy applications have concentrated on the location of parks on suitable locations outside urban conurbation. Important reasons for this are related not only to efficiency but also to ecological constraints. Visibility disturbance and the impact on the local natural environment seem to have become major issues causing a negative public attitude towards these systems. Local authorities should therefore put more efforts into investigating how wind energy applications can be integrated into urban planning on smaller scales.

Price facilitation is not sufficient to repay initial expenses

In many cases national authorities implement financial structures for wind energy initiatives as one of the instruments in their stimulation policies for energy innovations. However, even with subsidies, windmill parks almost never operate with positive results due to the high costs of the technology in comparison with the generation of energy. This raises the question as to what extent governments should give financial support to such local initiatives. Too much financial support would stimulate applications in sub-optimal locations where, even when systems could become more efficient in the future, initiatives would never operate profitably without financial help. Therefore, the national authorities should apply clear selection criteria in their supporting policies.

Sunk costs of plants are high but so are the initial investment costs for the connection to the electricity grid

The only way for the barrier of sunk costs to be overcome is for windmill systems to operate profitably long term. Producers of systems should increase their efforts in R&D activities in order to improve the profitability of these initiatives. The profitability of windmill parks also depends on the costs incurred by operators of the systems in connection to the electricity grid. Local authorities should therefore enforce a close cooperation between electricity grid and windmills operators - in the case that these are not be the same parties.

Windmill parks demand extensive use of land surface, making plant location costs high in general

The financial problem about the use of land can be partly solved by local land planning. When local authorities are the operators of windmill parks, of course there is no significant problem as the parks are frequently part of the overall land-use planning put into practice by them. When the operators are separate bodies, the local authorities are able to facilitate land prices. Much depends on the characteristics of the land occupied by windmill parks. Costs will be lower as the parks are located in remote areas. However, taking the social costs into consideration, in most cases these will be higher when the park is farther outside the urban boundaries and closer to nature.

Other energy sources have more competitive prices

The actual low level of competition between wind energy and conventional forms of energy supply is not unique for wind energy but applies to all innovative energy forms. There are two main reasons why this is not a major bottleneck in the implementation of wind energy systems. Firstly, in the near future it may be expected that the price of wind energy will decline when these systems become more mature. Furthermore, on the longer term it may be expected that prices of non-renewable energies will increase. Thus, suppliers should conduct a long term product strategy based on making aeolian resources more competitive.

Windmill systems create environmental/sociological problems related to birds, noise and electromagnetic interference

The wider social costs of land occupation by windmill parks cannot be covered by regular market-based land prices because of the impact of these parks on the local environment, in terms of visibility and noise disturbance and their impacts on animal life. Bird mortality is only an example of this. Producers should therefore include long term product strategies in their research. Integrating wind energy systems into the urban environment will reduce the impact on the natural environment, but the question is to which extent can this technology ultimately be absorbed into the urban environment. Probably a good solution would be for local authorities to place windmill parks in isolated areas, within reasonable proximity to cities but of minor importance from the point of view of threatened animals. Another option would be to make use of areas already intersected by traffic corridors, so that these systems have only marginal impact in terms of noise and visibility disturbances.

11.4 Solar Water Heating

Particular aesthetic problems may occur when the equipment is installed in the interior of the house

The actual implementation of solar water heating systems in their domestic appliances is already taking place on large scales in many cities. Therefore these systems have already provided much experience on aesthetic aspects, both inside and outside residences. In order to achieve a broader acceptance by the public, it is necessary to integrate the equipment in the interior so that the visibility impact is minimized. This would be a task not only for system producers but also for installers/housing constructors, operators/maintainers of the systems, and quite frequently, the local energy companies. Concerning the in-house equipment, the aesthetic aspect relates not only to shape and colour but also to the space occupied by these systems and their components. This becomes more important for smaller dwellings.

The acceptance process by the public may be slow

With a technological state which is actually quite mature, solar water heating is a product that lends itself to being marketed to individual households as the complete equipment is typically stand-alone for a residence. Modern marketing strategies may be adapted as a normal consumer product, with the difference that in most cases individual subsidies are granted with the purchase or rental of the equipment. One of the marketing strategies that is opted for, is based on a 'package' of innovative products. The acceptance by the public may be accelerated when the solar water heating product is offered in combination with, for instance, high efficiency central heating systems. This, however, requires cooperation between solar water heating suppliers, suppliers of other innovative energy products, local authorities and operators. Another, complementary, strategy would be product differentiation regarding e.g. placement, design and styling of the equipment.

In general the equipment produces more operating losses to the user than traditional sources do

Raising the positive public awareness to solar water heating by emphasizing its beneficial impact on the environment, will not sufficiently guarantee acceptance by the public. Without subsidies on the purchase of the equipment, there is not (yet) a financial break-even for the end-user in relation to his actual expenses on energy. Therefore it remains necessary to apply proper financial support schemes for individual households. In the case of new buildings, there would be an option to integrate the equipment into the new house, and into the price of the house. When no legal constraints occur, this would be a powerful method to extend implementation of solar water heating to larger scales in new building areas.

Solar water heating equipment is only economically attractive if more than one unit is installed collectively in one run

An efficient and reasonable way for solar water heating equipment to be installed, is collectively in a series of newly built houses. An important criteria is that the housing constructor is an expert in the installing of equipment. Such expertise frequently has been lacking in pilot projects and has even become a critical issue. When the equipment installer and the housing constructor are separate parties, it is of course very important that they cooperate well - both in the designing as well as in the installing phase. Also during renovation activities of a series of houses, it could be advantageous to install the equipment collectively.

There is a need for close cooperation between various actors at the supply end when solar water heating units are to be introduced to the public, collectively

A good organization for the planning and actual implementation of solar water heating systems is necessary to avoid (temporary) institutional obstacles during this process. Local authorities and energy companies must define tasks and divide responsibilities between the relevant supply actors. For instance, local authorities may impose the installation of solar water heating systems to land developers with the parcelling out of land, and to social housing corporations. On the other hand, energy companies or operators may carry out the technical coordination and maintenance tasks of solar water heating systems

11.5 CHP/District Heating

The implementation scale of district heating systems implies difficulties in terms of technical competence if the implementation has to be achieved short term, inter alia because of technical indivisibility
District heating systems have an inherently large implementation scale, hindering the experimentation with micro-level applications (on the level of dwellings). An extensive underground pipeline network is required. Therefore, the implementation of district heating has to be integrated with urban planning. Local authorities, suppliers and operators have to define a development plan with a continuous but manageable time path in order to maximize advantages from 'learning by doing' and avoid fundamental technical changes.

The technical compatibility between district heating and old technologies is generally good, except in the situation where methane is the old energy source. In this latter case a switch to district heating is difficult and costly
When a city's energy supply system uses methane as its source, energy companies and local authorities should pursue the coexistence between methane and district heating fed by other sources. This is easier when methane and district heating are not competitive, i.e. when the energy supply system is integrated.

Costs of information campaigns are high
The implementation of district heating networks takes time over long periods, as they are interwoven with urban expansion. Therefore, even with an existing successful application in a part of the city, local authorities and energy operators should not decrease their efforts in information campaigns. The cost of information campaigns can be offset in the future by a smoother expansion of the existing district heating network.

The relatively high fixed cost factor of district heating makes financial losses of the system very dependent on small fluctuations in heat demand
The main financial risk about implementing district heating systems is related to the relatively high fixed cost (of the underground infrastructure) in the overall costs. This fact makes the financial viability of these systems highly dependent on future variations in the demand for heat in the area served by the network. Many factors may determine a difference in demand in future. For instance, housing renovations with better insulation will automatically decrease the demand for heat. Operators of

the systems should take such risks into account by making proper estimations regarding future heating demands.

The construction of the necessary pipeline network is extremely labour intensive
There is a great advantage when urban district heating systems are constructed parallel to the building of new city quarters. This requires integrative urban planning in which the energy infrastructure is included. The total investment costs of district heating systems can be minimized in this way.

Opposition of local population to the location of the central plant
District heating systems require central plants located near to or in the city in order to be efficient. Although the combined heat power engines are relatively less polluting than conventional energy generation systems, this is often not popularly known, which could harm the positive public opinion on these systems. In order to raise public awareness, the operators of the systems should continue or expand their information campaigns among the local population to emphasize the advantages of district heating in terms of environmental benefits and efficiency gains.

The large size of district heating networks implies strong impacts on the liveability in the district covered.
The construction of district heating networks may have a large impact on the liveability of the covered area, due to the underground infrastructure. The disturbances may be work in progress on the streets, interventions on other networks and maintenance of the network. Disturbances during construction can be minimized when the network is constructed in newly built areas. In other cases, operators and local authorities should reduce the discomforts for citizens through compensations, and should provide detailed information on the time path of construction. This also applies to maintenance works.

11.6 Refuse Incineration

There is often a lack of specific know-how, while costs of R&D activities are high
Although the energy recovery from waste incineration is a technology that already has been implemented in many cities, there often still is a lack of specific know-how. System suppliers, waste incineration operators and local authorities have therefore to

stimulate acquisition of know-how through the enlargement of the scientific exchange network. In addition, national and local authorities should strengthen a system of incentives and 'soft loans' to support R&D activities.

There are problems related to safety because the composition of waste varies (there are no homogeneous contents)
National authorities should strengthen human security systems in waste incineration plants which recover energy, by using clear safety indicators, creating quality laws and increasing controls.

The recovery of energy from waste requires a link to an electricity or district heating network
An efficient operation of energy recovery from waste can be reached when the incineration plant is connected to the electricity grid or the local district heating network. This might vary from small-scale networks (e.g. connection to a few public buildings) to large-scale networks, depending on the capacity of the waste incineration plants. Local authorities should strengthen policies that favour incineration systems linked district heating systems since a relatively high level of efficiency is reached in this way.

External financing by means of subsidies is relatively difficult in comparison with other new energy technologies as waste incineration is not always perceived as an innovative concept
The technology of energy recovery from waste incineration is sometimes perceived to be less innovative than other renewable energy applications. Nevertheless, this technology is widely being implemented and existing implementations already make up a significant part of urban energy supply. In this stage of implementation it is very important that the applications become also financially viable. This is very difficult, as incineration plants usually also have to deal with tightened emission norms, and have to make - in addition to the investments in energy recovery - large investments in emission reduction systems. When external financing by means of subsidies is difficult to obtain, these initiatives could rely on third party financing programmes.

There is sometimes a weak connection with the local waste management system
The most efficient way of implementing energy recovery from waste incineration is by extending existing waste incineration plants. However, full integration with the local waste incineration system does not always exist and operators and local

authorities have to increase their efforts to develop an integrated waste management system.

The number of (potential) system operators is limited as a high degree of specialization is required
The technology of energy recovery from waste incineration often goes paired with complementary technologies for reducing emissions. Altogether there may be a complex new system, and often there is (still) a lack of specialized companies to operate it. Local authorities could make the market more competitive by stimulating proper training programmes for waste incineration operating companies.

The polluting emissions resulting from the incineration of waste and the opposition to the waste incineration activity by inhabitants of the surrounding area may be high

Operators have to use advanced pollution controlling technologies like depuration systems in order to minimize emissions from the waste incineration and the recovery of energy. At the same time, they should, together with local authorities, make the public aware (via information campaigns) of the environmental advantages stemming from a mass reduction of refuse and from a clean incineration (depuration system). Another possibility for local authorities is to create a compensation scheme for the people and firms damaged by the proximity of the incinerator, e.g. rental or house buying facilities or land reclamation.

11.7 Energy from Biogas

Sunk costs of powerful engines that fully exploit the biogas potential, are high
The problem of large sunk costs of investments in biogas energy recovery can form a serious obstacle to further implementation of this technology. Naturally to solve this problem, suppliers should increase research efforts in exploring ways for an optimal exploitation of biogas. But also these kinds of applications require active participation from the public to become financially viable, and to overcome sunk costs. By means of waste collection management, national and local governments should extend the presence of biogas, as far as possible, or enforce separate collection of organic waste and other waste.

In situations where two or more responsible actors exist (e.g. waste incineration operators and energy companies), a lack of technical competence can occur
Just as in the case of the recovery of energy from waste incineration, from a technical and managerial point of view it would be most efficient to organize responsibilities in such a way that the actor who will operate the energy recovery from biogas, is skilled for this purpose. The best possibility is an integrated company that manages both refuse and also the energy supply to the city. Local authorities should implement such integration in their urban energy policies.

Applications are not (yet) profiting and existing subsidies may be limited in the future
National authorities should stimulate the financial independence of initiatives in biogas energy recovery by tightening the existing subsidy structure and limiting automatic prolongation of this financing until the initiatives are able to carry on independently. They should also map initiatives which can operate on the basis of subsidies in order to understand the associated risk of interruptions.

There is frequent opposition from the local population to activities related with the recovering of biogas
Suppliers and local authorities should make the public aware of the advantages stemming from this kind of initiative through campaigns explaining the measures taken to reduce dangers related to waste dumping and bad smell.

11.8 Outlook

In the context of sustainable city policy, it should be recognized that cities are not passive spatial units victimized by anonymous global environmental developments, but may play an active role in producing sustainable development in a multiplicity of relevant fields, such as housing, employment or environmental quality. This promising profile of the city as a window of environmental opportunities is increasingly coming to the fore in institutional agenda formation for sustainable development, not only for the developed - but also for the developing world. This means that urban development planning has to address a wide variety of issues and objectives regarding the economy, the environment, energy supplies and the socio-economic distribution of costs and benefits. Clearly, there is no uniform panacea for sustainable urban planning because there are site-specific, environmental, economic, political and socio-cultural factors to be considered. Decision-makers in both the public and the business sector play a critical role in all aspects of urban life, since a sustainable city has to be created by people themselves.

Strategies favouring environmentally sustainable city development, and in particular energy demand and supply, may be varied in nature, ranging from the introduction of advanced new technologies to information campaigns, financial incentives and strict regulations. Such strategies which may, among other things, concern economic sectors, social groups or land use in the city, should of course be consistent with sustainability policies elsewhere (adjacent areas, or higher geographical levels) up to the level of global environmental policy-making. In view of the threats caused by spatial environmental spill-overs to other areas (a phenomenon also witnessed by concepts like weak/strong sustainability or external/internal sustainability), coordinated action is needed to prevent spatial imbalances in sustainability policies at all levels and in all areas of interconnected spatial systems.

It is thus clear that sustainable city policy requires a multi-faceted strategy, in which socio-economic interests are brought into harmony with environmental interests. Such ideas have already been advocated more than a decade ago by Lynch (1981) who claimed: "*The good city is one in which the continuity of this complex ecology is maintained while progressive change is permitted*". Lynch suggests five dimensions for judging such an urban quality: vitality, sense, fit, access and control. It goes without saying that the empirical application of such principles will require much applied research on sustainability conditions in a variety of urban settings.

In this context, the position of Third World cities deserves more attention. Cities in the developing world often face a serious struggle for continuity and survival on the basis of short-run economic interests. Even though it is widely recognized that favourable environmental quality conditions are a sine qua non for long-term sustainability, severe short-term survival behaviour prevents many cities from developing a co-evolutionary strategy in which economic interests are brought in harmony with urban ecology. Poverty situations are apparently playing a more privileged role than the strife for environmental quality improvement.

Nevertheless, the dilemma between micro rationality and collective sustainability policy ought to be considered also from the opposite perspective. Low environmental quality generates a multiplicity of negative stimuli for human health, education and social welfare and is hence detrimental to the welfare position of the economically less privileged, and in the end, is at odds with a sustainable urban habitat.

This observation justifies strict public environmental policies in Third World cities, as investments in these sectors will pay off in economic terms in the long run. In this respect, there is no real conflict of interest between urban economic development and urban environmental sustainability.

References

Antonelli, C. (1995), *The Economics of Localised Technological Change and Industrial Dynamics*, Kluwer, Dordrecht.

Archibugi, F. (1997), *The Ecological City and the City Effect*, Ashgate, Aldershot, UK.

Archibugi, F. and P. Nijkamp (eds.) (1989), *Economy and Ecology: Towards Sustainable Development*, Kluwer, Dordrecht.

Arthur, W.B. (1988), Competing Technologies: an Overview, *Review of Economic Studies*, no.29, 155-173.

Arthur, W.B. (1990), Positive Feed-backs in the Economy, *Scientific American*, February, 80-85.

Atkinson, A. (1994), The Contribution of Cities to Sustainability, *Third World Planning Review*, vol.16, no.2, 97-103.

Baayens, S. and P. Nijkamp (1998), Meta-analytic Methods for Explanatory and Exploratory Policy Research, *Journal of Policy Modelling* (forthcoming).

Barrow, C.J. (1995), Sustainable Development, Concept, Value and Practice, *Third World Planning Review*, vol.17, no.4, 369-386.

Baumol, W.J. and W.E. Oates (1988), *The Theory of Environmental Policy*, Cambridge University Press, Cambridge.

Bergh, J.C.J.M. van den (1991), *Dynamic Models for Sustainable Development*, Thesis Publishers, Amsterdam.

Bergh, J.C.J.M. van den (1996), *Ecological Economics and Sustainable Development: Theory, Methods and Applications*, Edward Elgar, Cheltenham.

Bergh, J.C.J.M. van den, K. Button, P. Nijkamp and G. Pepping (1997), *Meta-analysis in Environmental Economics*, Kluwer, Dordrecht.

Bergh, J.C.J.M. van den and J. van der Straaten (eds.) (1994), *Toward Sustainable Development Concepts, Methods and Policy*, Island Press, Washington, DC.

Breheny, M. (1993), The Compact City, *Built Environment*, vol.18, no.4, 241-246.

Breheny, M.J. (ed.) (1994), *Sustainable Development and Urban Form*, Pion, London.

Bressan, B. (1996), Il Vento di Montagna, *Energia e Materie Prime*, June, 44-47.

Button, K.J. (1992), Transport Regulation and the Environment in Low Income Countries, *Utilities Policy*, July, 248-257.

Button, K.J. and J.C.J.M. van den Bergh (1997), Meta-analysis of Environmental Issues in Regional, Urban and Transport Economics, *Urban Studies*, vol.34, 927-944.

Button, K.J. and J. Kerr (1996), The Effectiveness of Traffic Restraint Policies: a Simple Meta-regression Analysis, *International Journal of Transport Economics*, vol.23, no.2, 213-225.

Button, K.J. and P. Nijkamp (1997), Environmental Policy Assessment and the Usefulness of Meta-analysis, *Socio-Economic Planning Sciences*, vol.31, no.3, 231-240.

Button, K.J. and D.W. Pearce (1989), Improving the Urban Environment: How to Adjust National and Local Government Policy for Sustainable Urban Growth, *Progress in Planning*, vol.32, no.3, 137-184.

Button, K.J. and W. Rothengatter (1993), Global Environmental Degradation: The Role of Transport, *Transport, the Environment and Sustainable Development* (D. Banister and K.J. Button, eds.), E&FN Spon, London, 161-178.

Camagni, R. (1985), Spatial Diffusion of Pervasive Process Innovation, *Papers of the Regional Science Association*, vol.58, 83-96.

Camagni, R. (1996), Lo Sviluppo Urbano Sostenibile: le Ragionie e i Fondamenti di un Programma di Ricerca, *Economia e Pianificazione della Città Sostenibile* (R. Camagni, a cura di), Il Mulino, 13-51.

Camagni, R. (1998), Sustainable Urban Development: Definition and Reasons for a Research Programme, *International Journal of Environment and Pollution*, vol.10, no.1, 6-27.

Camagni, R. and R. Capello (1991), Le Caratteristiche delle Nuove Tecnologie di Comunicazione e loro Interazione con la Domanda, *Computer Network: Mercati e Prospettive delle Tecnologie di Comunicazione* (R. Camagni, ed.), Etas Libri, Milan, 3-46.

Camagni, R. and R. Capello (1998), Indivisibilità e Irreversibilità nei Processi di Adozione di Tecnologie Rinnovabili in Ambiente Urbano, *I Luoghi della Trasformazione e dell' Innovazione* (F. Boscacci, F. and L. Senn, eds.), SEAT, Turin, 223-246.

Camagni, R. and R. Cappellin (1984), Cambiamento Strutturale e Dinamica della Produttività nelle Regioni Europee, *Cambiamento Tecnologico e Diffusione Territoriale*, (R. Camagni, R. Cappellin and G. Garofoli, eds.), Franco Angeli, Milan, 131-337.

Camagni, R., R. Capello and P. Nijkamp (1997), Transport and Communication for Urban Sustainable Development, *Transport and Communications for Sustainable Development* (HABITAT), HABITAT, Nairobi.

Camagni, R., R. Capello and P. Nijkamp (1998), Towards Sustainable City Policy an Economy, Environmental, Technology Nexus, *Ecological Economics,* vol.24, no.1, 103-118.

Capello, R. (1998), Urban Returns to Scale and Environmental Resources: an Estimate of Environmental Externalities in an Urban Production Function, *International Journal of Environment and Pollution*, vol.10, no.1, 28-45.

Capello, R. and P. Nijkamp (eds.) (1998), Urban Sustainable Development, *Special Issue of the International Journal of Environment and Pollution*, vol.10, no.1.

Capello, R. and H. Williams (1992), Computer Network Trajectories and Organisational Dynamics: a Cross-national Review, *The Economics of Information Networks* (C. Antonelli, ed.), North Holland, Amsterdam, 347-362.

Cavaggioni, M. (1992), Produzione di Energia da Fonti Rinnovabili, *Energia e Innovazione*, ENEA, no.2-3, 43-49.

Cloke, P.J. and C.C. Park (1985), Integrated Management Strategies, *Rural Resource Management*, Croom Helm, Andover, 408-448.

Coiante, D. (1992), Le Nuove Fonti Rinnovabili, Strategie e Prospettive, *Energia e Innovazione*, ENEA, no.4-5, 33-39.

Coiante, D. (1996), Le Barriere per la Diffusione su Larga Scala dell'Energia Fotovoltaica, *Energia e Materie Prime*, June, 10-15.

Corvi, M. (1996), Il Vento dell'UE, *Energia e Materie Prime*, June, 62-72.

Daly, H.E. (1991), *Steady State Economics*, Island Press, Washington, D.C.

David, P. (1985), Clio and the Economics of Qwerty, *AEA Papers and Proceedings*, vol.75, no.2, 332-337.

David, P. (1992), Information Network Economics: Externalities, Innovation and Evolution, *The Economics of Information Networks* (C. Antonelli, ed.), North Holland, Amsterdam, 103-106.

David, P. and S. Greenstein (1990), The Economics of Compatibility Standards: an Introduction to Recent Research, *Economics of Innovation and New Technology*, no.1, 3-42.

Davidson, J. (1988), *How Green is Your City?*, Bedford Square Press, London.

De Paoli, L. (1995), Privatizzazione del Sistema Elettrico e Nuovi Rapporti tra Enel e Minicipalizzate, *Economia delle Fonti di Energia e dell'Ambiente*, no.3, 75-92.

Dosi, G. (1982), Technological Paradigms and Technological Trajectories: a Suggested Interpretation of the Determinants and Direction of Technological Change, *Research Policy*, vol.11, no.3, 147-162.

Dosi, G. (1988), Sources, Procedures and Microeconomic Effects of Innovation, *Journal of Economic Literature*, vol.XXVI, 1120-1171.

Dosi, G., C. Freeman, R. Nelson, G. Silverberg and L. Soete (eds.) (1988), *Technical Change and Economic Theory*, Pinter, London.

Elkin, T., D. MCLaren and M. Hillman (1991), *Reviving the City: Towards Sustainable Urban Development*, Friends of the Earth, London.

Environmental Protection Agency (EPA) (1980), *Emission Reduction Banking Manual*, EPA, Washington.

Erdmann, G. (1993), Evolutionary Economics as an Approach to Environmental Problems (H. Giersch, ed.), *Economic Progress and Environmental Concern*, Springer-Verlag, Berlin, 61-83.

EU (1996), *Energy for the Future: Green Paper of Energy*, Brussels.

European Commission (1990), *GreenPaper on the Urban Environment*, EC, Brussels.

Freeman, C. (1987), *Technology Policy and Economic Performance*, Frances Pinter, London.

Fusco Girard, L. and P. Nijkamp (1997), *Le Valutazioni per lo Sviluppo Sostenibile della Città e del Territorio*, Franco Angeli, Milan.

Gaffard, J-L. (1990), *Economie Industrielle et de l'Innovation*, Dalloz, Paris.

Garbati, R. and P. Galliano (1993), *Valutazione del Potenziale Eolico della Valle Dora Riparia*, AEM, Turin.

Geenhuizen, M.S. van, and P. Nijkamp (1997), Sustainable Cities: Challenges of an Integrated Planning Approach, *Essay on Regional Science* (C.R. Pathak, ed.), Regional Science Association India, Calcutta, 1997, 3-21.

Giaoutzi, M. and P. Nijkamp (1994), *Decision Support Models for Regional Sustainable Development*, Avebury, Aldershot.

Gibbs, D. (1994), Towards the Sustainable City, *Town Planning Review*, vol.65, no.1, 99-109.

Girardet, H. (1992a), *The Gaia Atlas of Cities*, Gaia Books, London.

Girardet, H. (1992b), *Cities: New Directions for Sustainable Urban Living*, Gaia Books, London.

Greco, S., B. Matarazzo and R. Slowinski (1995), Rough Set Approach to Multi- attribute Choice and Ranking Problems, ICS Research Report 38/95, Warsaw University of Technology, Warsaw.

Hahn, R.W. (1984), Market Power an Transferable Property Rights, *Quarterly Journal of Economics*, no.99, 753-765.

Haughton, G. and C. Hunter (1994), *Sustainable Cities*, Jessica Kingley, London.

Hedges, L.V. and I. Olkin (1985), *Statistical Methods for Meta-analysis*, Academic Press, New York.

Henderson, H. (ed.) (1981), *The Politics of the Solar Age*, Archor Books, New York.

Howe, C.W. (1994), Taxes versus Tradeable Discharge Permits: A Review in the Light of the US and European Experience, *Environmental and Resource Economics*, vol.4, 151-169.

Hunter, J.E., F.L. Schmidt and G. Jackson (1982), *Advanced Meta-analysis: Quantitative Methods for Cumulating Research Findings Across Studies*, Sage, Beverly Hills.

Hunter, C. (1998), Perceptions of the Sustainable City and Implications for Frech Water Resources Management, *International Journal of Environment and Pollution*, vol.10, no.1 84-103.

IEA (1987), *Renewable Sources of Energy*, OECD, Paris.

IEA (1991), *Energy Efficiency and the Environment*, OECD, Paris.

IEA (1994a), *Energy Balances of OECD Countries 1991-1992*, OECD, Paris.

IEA (1994b), *District Heating and Combined Heat and Power Systems*, OECD, Paris.

James, D.E., P. Nijkamp and J.B. Opschoor (1989), Ecological Sustainability and Economic Development, *Economy and Ecology* (F. Archibugi and P. Nijkamp, eds.), Kluwer, Dordrecht, 21-42.

Jenks, M., V. Williams and E. Wilson (eds.) (1996), *The Compact City: a Sustainable Urban Form?*, E & F Spon, London.

Katz, M. and C. Shapiro (1992), Product Introduction with Network Externalities, *The Journal of Industrial Economics*, vol.XL, no.1, 55-83.

Katz, M. and C. Shapiro (1985), Network Externalities, Competition and Compatibility, *The American Economic Review*, vol.75, no.3, 424-440.

Katz, M. and C. Shapiro (1986), Technology Adoption in the Presence of Network Externalities, *Journal of Political Economy*, 822-841.

Klaassen, G. (1996), *Acid Rain and Environmental Degradation*, Edward Elgar, Cheltenham.

Koutstaal, P. (1997), *Energy Policy and Climate Change*, Edward Elgar, Cheltenham.

Liebowitz, S. and S. Margolis (1990), The Fable of the Keys, *Journal of Law and Economics*, vol.33, 1-25.

Light, R.J. and D.B. Pillemer (1984), *Summing Up - The Science of Reviewing Research*, Harvard University Press, Cambridge MA.

Lindley, D. (1995), Wind Energy: Status and Prospects, *Proceedings of the Conference on Renewable Energy Developent*, November, Venice, 481-491.

Lorenzoni, A. (1994), Prospettive Economiche per le Energie Rinnovabili nel Settore Elettrico: l'Eolico ed il Fotovoltaico in Italia, *Economia delle Fonti d'Energia*, no.1, 45-107.

Lundsager, P. (1995), Integration of Renewable Energy into Combined Heat and/or Local Power Supply, *Proceedings of the Conference on Renewable Energy Development*, November, Venice, 469-480.

Lynch, K. (1981), *A Theory of Good City Form*, MIT Press, Cambridge, Mass.

Macintosh, D.R. (1973), *The Economy of Airborne Emissions: The Case of an Air Rights Market*, Praeger Publishers, New York.

Matarazzo, B. and P. Nijkamp (1997), Meta-analysis for Comparative Environmental Case Studies: Methodological Issues, *International Journal of Social Economics*, vol.34, no.7-9, 799-811.

MEZ (Dutch Ministry of Economic Affairs) (1992), *Verhandelbare CO_2 Emissierechten in Nederland en de EG*, The Hague.

Montgomery, D.W. (1972), Markets in Licences and Efficient Pollution Control Programs, *Journal of Economic Theory*, no.5, December, 395-418.

Nelson, R. and S. Winter (1977), In Search of a Useful Theory of Innovation, *Research Policy*, vol.6, 36-76.

Nelson, R. and S. Winter (1982), *An Evolutionary Theory of Economic Changes*, Harvard University Press, Cambridge, Mass.

Newman, P. and J. Kenworthy (1989), *Cities and Automobile Dependence*, Gower, Aldershot, UK.

Nijkamp, P. (ed.) (1991), *Urban Sustainability*, Gower, Aldershot, UK.

Nijkamp, P. (1996), Improving Urban Environmental Quality: Socioeconomics Possibilities and Limits, *Urban Poverty in Asia, a Survey of Critical Issues* (E.M. Pernia, ed.), Oxford University Press, Hong Kong, 241-292.

Nijkamp, P., P. Lasschuit and F. Soeteman (1992), Sustainable Development in a Regional System, in Breheny, M. (ed.), *Sustainable Development Planning*, Pion, London, 112-136.

Nijkamp, P. and J.B. Opschoor (1997), Urban Environmental Sustainability: Critical Issues and Policy Measures in a Third-World Context, *Regional Science in Developing Countries* (M. Chatterji and Y. Kaizhong, eds.), MacMillan, London, 52-73.

Nijkamp, P. and J. Ouwersloot (1998), A Decision Support system for Regional Sustainable Development, *Models for Sustainable Development* (J.C.J.M. van den Bergh and M. Hofkes, eds.), Edward Elgar, Cheltenham (forthcoming).

Nijkamp, P. and G. Pepping (1998), A Meta-Analytic Evaluation of Sustainable City Initiatives, *Urban Studies*, vol.35, 1481-1500.

Nijkamp, P. and A.H. Perrels (1994), *Sustainable Cities in Europe*, Earthscan, London.

Nijkamp, P. and Th. Ursem (1998a), Market solutions for Sustainable Cities, *International Journal of Environment and Pollution*, vol.10, no.1, 46-83.

Nijkamp, P. and M. Ursem (1998b), Critical Success Factors for Urban Energy Policies: A Meta-analysis, Research paper, Department of Regional Economics, Free University, Amsterdam.

Nijkamp, P., J. Vleugel, R. Maggi and I. Masser (1994), *Missing Transport Networks in Europe*, Avebury, Aldershot, UK.

OECD (1978), *Urban Environmental Indicators*, Paris.

OECD (1990), *Environmental Policies for Cities in the 1990's*, Paris.

OECD (1995), *Urban Energy Handbook*, OECD, Paris.

Orishi, T. (1994), A Capacity Approach for Sustainable Urban Development, *Regional Studies*, vol.28, no.1, 39-51.

Opschoor, J.B. (1992), Sustainable Development, the Economic Process and Economic Analysis, *Environment, Economy and Sustainable Development* (J.B. Opschoor, ed.), Wolters Noordhoff, Groningen, 1-23.

Orishimo, I. (1982), *Urbanisation and Environmental Quality*, Kluwer, Dordrecht.

Owens, S. (1986), *Energy, Planning and Urban Form*, Pion, London.

Owens, S. and P. Rickaby (1992), Settlements and Energy Revisited, *Built Environment*, vol.18, no.4, 247-252.

Pawlak, Z. (1982), Rough sets, *International Journal of Information and Computer Sciences*, vol.11, no.5, 341-356.

Pawlak, Z. (1991), *Rough Sets. Theoretical Aspects of Reasoning About Data*, Kluwer, Dordrecht.

Pearce, D., A. Markandya and E.B. Barbier (1989), *Blueprint for a Green Economy*, Earthscan, London.

Pelt, M.J.F. van, A. Kuyvenhoven and P. Nijkamp (1990), Project Appraisal and Sustainability: Methodological Challenges, *Project Appraisal*, 5(3), 139-158.

Pezzy, J. (1989), *Economic Analysis of Sustainable Growth and Sustainable Development*, WP15, Environment Department, The World Bank, Washington, D.C.

Poll, R. van (1997), *The Perceived Quality of the Urban Residential Environment*, Ph.D. Thesis, State University, Groningen.

Porter, M. (1990), *The Competitive Advantage of Nations*, Free Press, New York.

Rees, W.E. (1992), Ecological Footprints and Appropriated Carrying Capacity, *Environment and Urbanisation*, vol.4, 121-130.

Rosdorff, S., L.K. Slager, V.M. Sol and K.F. van der Woerd (1994), *De Stadsstolp: Meer Ruimte voor Milieu én Economie*, Institute for Environmental Studies, Free University, Amsterdam.

Rosenberg, N. (1982), *Inside the Black Box: Technology and Economics*, Cambridge University Press, Cambridge.

Rosenthal, R. (1991), *Meta-Analytic Procedures for Social Research*, Sage, Beverly Hills.

Scherer, F.M. (1980), *Industrial Market Structure and Economic Performance*, Rand McNally, Chicago.

Selman, P. (1996), *Local Sustainability*, Paul Chapman, London.

Sjunnesson, L. (1995), Energy from Biomass, *Proceedings of the Conference on Renewable Energy Development*, November, Venice, 497-502.

Slowinski, R. and J. Stefanowski (1994), Handling Various Types of Uncertainty in the Rough Set Approach, *Rough Sets, Fuzzy Sets and Knowledge Discovery* (W.P. Ziarko, ed.), Springer, Berlin, 191-218.

Stren, R., R. White and J. Whitney (eds.) (1995), *Sustainable Cities*, Westview Press, Oxford.

Struyk, R.J., M.L. Hoffman and H.M. Katsura (1990), *The Market for Shelter in Indonesian Cities*, The Urban Institute Press, Washington, D.C.

Stuurgroep Integraal Milieubeleidsplan Energiedistributiesector (1991), *Milieu Actie Plan van de Energiedistributiesector 1991*, Stuurgroep Integraal Milieubeleidsplan Energiedistributiesector, Arnhem.

Theil, H. (1971), *Principles of Econometrics*, John Wiley, New York.

Tietenberg, T.H. (1980), Transferable Discharge Permits and the Control of Stationary Source Air Pollution: A Survey and Synthesis, *Land Economics*, vol.56, no.4, 391-416.

Turner, R.K., D. Pearce and I. Bateman (1994), *Environmental Economics: An Elementary Introduction*, Harvester Wheatsheaf, Hemel Hempstead, Hertfordshire, UK.

Tweede Kamer der Staten-Generaal (1995a), *Derde Energienota*, SDU Uitgeverij, The Hague.

Tweede Kamer der Staten-Generaal (1995b), *Energiebesparingsbeleid*, SDU Uitgeverij, The Hague.

United Nations Conference on Environment and Development (1992), *Agenda 21 - Action Plan for the Next Century*, UNCED, Rio de Janeiro.

Verhoef, E., P. Nijkamp and P. Rietveld (1997), Tradeable Permits: Their Potential in the Regulation of Road Transport Externalities, *Environment & Planning B*, vol.24, 527-548.

Vleugel, J. (1995), *De Mileugebruiksruimte voor Duurzaam Verkeer en Vervoer*, Delftse Universitaire Press, Delft.

Wackernagel, S. and W.E. Rees (1996), *Our Ecological Footprint*, New Society Publ., Gabriola Island, B.C.

WCED (1987), *Our Common Future* (Brundtland Report), World Commission on Environment and Development, Oxford University Press, Oxford.

White, R.R. (1994), *Urban Environmental Management*, John Wiley, New York.

Whittaker, S. (1995), *An International Guide to Local Agenda 21*, HMSO, London.

Wolf, F. (1986), *Meta-Analysis: Quantitative Methods for Research Synthesis*, Sage, Beverly Hills.

Yanarella, E. and R.S. Levine (1993), The Sustainable Cities Manifesto, *Built Environment*, vol.18, no.4, 301-313.

Yogev, A. and A. Kribus (1996), Large Solar Thermal Power Plants, *Proceedings of the Conference on Renewable Energy Development*, November, Venice, 521-531.

Note

Though the volume is the integrated result of a joint effort of three authors, R. Capello has written Chapters 4, 5, 6 and 7, while the other two authors are mainly responsible for the remaining chapters.

Subject Index

Lightning Source UK Ltd.
Milton Keynes UK
UKOW020947291011

181132UK00001B/25/A

9 783540 648055